Jakob Abermann

Glaciers in Austria

Jakob Abermann

Glaciers in Austria

past and present

Südwestdeutscher Verlag für Hochschulschriften

Impressum/Imprint (nur für Deutschland/only for Germany)
Bibliografische Information der Deutschen Nationalbibliothek: Die Deutsche Nationalbibliothek verzeichnet diese Publikation in der Deutschen Nationalbibliografie; detaillierte bibliografische Daten sind im Internet über http://dnb.d-nb.de abrufbar.
Alle in diesem Buch genannten Marken und Produktnamen unterliegen warenzeichen-, marken- oder patentrechtlichem Schutz bzw. sind Warenzeichen oder eingetragene Warenzeichen der jeweiligen Inhaber. Die Wiedergabe von Marken, Produktnamen, Gebrauchsnamen, Handelsnamen, Warenbezeichnungen u.s.w. in diesem Werk berechtigt auch ohne besondere Kennzeichnung nicht zu der Annahme, dass solche Namen im Sinne der Warenzeichen- und Markenschutzgesetzgebung als frei zu betrachten wären und daher von jedermann benutzt werden dürften.

Coverbild: www.ingimage.com

Verlag: Südwestdeutscher Verlag für Hochschulschriften GmbH & Co. KG
Dudweiler Landstr. 99, 66123 Saarbrücken, Deutschland
Telefon +49 681 37 20 271-1, Telefax +49 681 37 20 271-0
Email: info@svh-verlag.de

Approved by: Innsbruck, Leopold-Franzens Universität, Diss., 2011

Herstellung in Deutschland:
Schaltungsdienst Lange o.H.G., Berlin
Books on Demand GmbH, Norderstedt
Reha GmbH, Saarbrücken
Amazon Distribution GmbH, Leipzig
ISBN: 978-3-8381-2881-8

Imprint (only for USA, GB)
Bibliographic information published by the Deutsche Nationalbibliothek: The Deutsche Nationalbibliothek lists this publication in the Deutsche Nationalbibliografie; detailed bibliographic data are available in the Internet at http://dnb.d-nb.de.
Any brand names and product names mentioned in this book are subject to trademark, brand or patent protection and are trademarks or registered trademarks of their respective holders. The use of brand names, product names, common names, trade names, product descriptions etc. even without a particular marking in this works is in no way to be construed to mean that such names may be regarded as unrestricted in respect of trademark and brand protection legislation and could thus be used by anyone.

Cover image: www.ingimage.com

Publisher: Südwestdeutscher Verlag für Hochschulschriften GmbH & Co. KG
Dudweiler Landstr. 99, 66123 Saarbrücken, Germany
Phone +49 681 37 20 271-1, Fax +49 681 37 20 271-0
Email: info@svh-verlag.de

Printed in the U.S.A.
Printed in the U.K. by (see last page)
ISBN: 978-3-8381-2881-8

Copyright © 2011 by the author and Südwestdeutscher Verlag für Hochschulschriften GmbH & Co. KG and licensors
All rights reserved. Saarbrücken 2011

TABLE OF CONTENTS

Abstract 3

Zusammenfassung 4

1. Introduction 6

2. Glacier and climate change 8

    2.1. Before the pleistocene 8

    2.2. Ice ages 10

    2.3. The Last Glacial Maximum 11

    2.4. Deglaciation History 13

    2.5. The Holocene 14

    2.6. The Little Ice Age 17

    2.7. After the Little Ice Age until 1969 21

    2.8. The inventories 1969 and 1998 22

3. Glaciers in Austria 25

    3.1. Types 25

        3.1.1. Valley glaciers 25

        3.1.2. Cirque glaciers 26

        3.1.3. Regenerated glaciers 27

        3.1.4. Hanging glaciers 27

        3.1.5. Plateau-like glaciers 27

    3.2. Glacier distribution, glacier changes and topography 27

    3.3. Glacier distribution, glacier changes and climate 33

4. Very recent glacier changes 34

5. Regional considerations 35

6. Glacier inventories: a climatic interpretation 36

    6.1. Introduction to mass balance determination 36

6.1.1. Direct glaciological method 36

6.1.2. Geodetic method 36

6.1.3. Hydrologic method 37

6.2. Introduction to mass balance modelling 37

6.2.1. Temperature-index 38

6.2.2. Energy balance 38

6.3. Calibrating mass balance with GI-derived volume changes 39

7. Conclusions and Outlook 40

Appendix A: Publications relevant to the thesis 42

A1: Paper I: Climatic controls of glacier distribution and glacier changes in Austria 42

A2: Paper II: On the Potential of very high-resolution DEMs in glacial and periglacial environments 51

A3: Paper III: Quantifying changes and trends in glacier Area and Volume in the Austrian Ötztal Alps (1969-1997-2006) 65

A4: Poster I: Towards a third Austrian glacier Inventory: First results and a climatic interpretation 77

A5: Poster II: Synchroneous glacier retreat North and South of a central-Alpine Mountain divide 79

A6: Paper IV: A reconstruction of annual mass balances of Austria's glaciers from 1969 to 1998 81

Appendix B: Glacier changes - Documentary work 89

References 100

Acknowledgements 110

Curriculum Vitae 111

List of Publications and Presentations 114

Abstract

The glaciers in the Austrian Alps, their changes and their relationship to climate change are investigated. In Austria, there are about 900 glaciers, covering less than 450 km² in elevations between 2100 m and 3800 m. Two complete glacier inventories from 1969 and 1998 are used to present the glacier distribution and glacier changes at the end of the last century. A quantifiable relationship between glacier size, elevation and mean climatic values was found. Glacier changes between 1969 and 1998 were in total negative, interrupted by an advance in the late 1970s and early 1980s. In the past decade, glacier retreat accelerated. Airborne Laser Scanning (ALS, LIDAR) proves to be an ideal tool to monitor glacier area and volume changes accurately. For the Ötztal Alps, glacier recession between 1998 and a new LIDAR-derived inventory of 2006 is quantified: Area reduced by 8.2%, volume by 1.0 km³ and mean thickness by 8.2 m. The three glacier inventories provide the basis to compare rates of glacier recession for the two periods investigated. Rates of volume changes have increased more than rates of area changes; rates of changes of large glaciers have increased more in recent years than rates of changes of small glaciers did. There is evidence for a North-to-South-gradient of magnitudes of glacier changes: Stronger recession is observed in the Southern part. A model is applied to reconstruct annual balances for a large sample of glaciers in Austria (96% of total glacier covered area). The course of the annual balances is reproduced well by the model; a temporally adjusted set of tuning parameters improves its performance of the model, possible reasons for that are discussed.

ZUSAMMENFASSUNG

In dieser Arbeit werden die Gletscher der österreichischen Alpen, ihre Änderungen und ihr Verhältnis zu Klimaänderungen behandelt. Es gibt in Österreich ungefähr 900 Gletscher auf einer Fläche von knapp 450 km² in Höhenlagen zwischen 2100 m und 3800 m. Zwei vollständige Inventare aus den Jahren 1969 und 1998 dienen als Grundlage, um die Verbreitung der österreichischen Gletscher und deren Änderungen am Ende des letzten Jahrhunderts zu zeigen. Es wurde eine Beziehung zwischen der Größe, der Höhenlage und mittleren klimatischen Verhältnissen eines Gletschers gefunden und quantifiziert. Gletscheränderungen waren in Summe negativ zwischen 1969 und 1998, unterbrochen durch einen Vorstoß in den ausgehenden 1970er- und frühen 1980er-Jahren. Der Gletscherrückgang hat sich in der vergangenen Dekade beschleunigt, was mit Airborne-Laser-Scanning-Daten (LIDAR) untersucht wird. Diese Daten stellen sich als ideal heraus, um Gletscherflächen- und -volumsänderungen genau dokumentieren zu können. Die Gletscher der Ötztaler Alpen sind zwischen 1998 und dem neuen Inventar von 2006 um 8.2% kleiner geworden und das Volumen hat sich um 1.0 km³ reduziert, was einer mittleren Dickenänderung von -8.2 m entspricht. Die drei Inventare bieten die Grundlage um Änderungsraten zwischen zwei Perioden zu vergleichen: Änderungsraten der Volumina haben sich gegenüber Flächenänderungsraten deutlicher vergrößert; große Gletscher weisen generell eine stärkere Zunahme von Änderungsraten auf als kleine Gletscher. Ein Nord-Süd-Gradient der Flächen- und Volumsänderungen ist feststellbar mit stärkeren Rückgängen im Süden. Schließlich wird ein Modell angewandt, um für den überwiegenden Anteil der vergletscherten Fläche in Österreich (96% der gesamten Gletscherfläche) Jahresbilanzen zu rekonstruieren. Der zeitliche Verlauf der Bilanzen ist gut reproduzierbar; das Ergebnis wird verbessert, indem sich zeitlich ändernde Eichfaktoren eingesetzt werden. Mögliche Ursachen dafür werden diskutiert.

Photo: W. Flaig, 1938. *Gletscherbuch*, Brockhaus, Leipzig, Germany, 196pp.

1. INTRODUCTION

Glaciers undergo significant changes in direct response to climate fluctuations. Our knowledge of global glacier fluctuations dates back far into pre-human times and temporal as well as spatial resolution improves with time. In recent years, the public perception of the significance and importance of understanding glacier changes has increased and a tremendous scientific effort has been made in order to improve knowledge of the major drivers, magnitudes, temporal and spatial variabilities and consequences of past and ongoing climate change on the cryosphere (e.g. Lemke and others (2007)).

Changes of the Alpine cryosphere are documented particularly well and their recent recession is of importance for different reasons: The main reason is, that by improving the understanding of magnitudes and spatial patterns of glacier recession and their causes, fundamental knowledge of mountain glaciers' reaction to climate change can be achieved. Locally, this knowledge is used and sought for by the economy: Hydropower companies are interested in the timing and amount of glacier water runoff within a year (Kuhn, 2003, Lambrecht and Mayer, 2009). Glacier ski resorts use information on the current distribution of ice and its recent rates of changes over decades (e.g. Fischer and others (2010)). Globally, we are convinced that the knowledge gained in a managable study area such as in the Alps, can be of benefit for more globally directed studies such as sea-level rise predictions or future glacier extent. Furthermore, it is the reliable long-term meteorological data that makes the Alps a unique and ideal study area to investigate the reliability of models under changed conditions.

This thesis thus aims at presenting glacier changes and their connection to climate change in the Austrian Alps with special attention directed towards the last 4 decades. It consists of four peer-reviewed papers in ISI-indexed journals and two posters presented at international conferences, which are all reproduced in Appendix A. It is structured as follows: In Chapter 2 a brief general descriptive review on glacier and climate change in the Earth's history is given with special emphasis on the Eastern Alps. This is done introductorily to bring the magnitude of the more recent glacier changes into a more comprehensive context. Glacier changes until the second complete glacier inventory (1998) are reviewed in this chapter. Chapter 3 then describes extent, morphology, location and size distribution of the Austrian glaciers in general. Furthermore the connection between the glacier distribution and the general climatic setting as well as a qualitative relation between the glacier changes and climate change is assessed. Chapter 4 presents a methodological part that shows the potential of airborne laser scanning (ALS, LIDAR) as an apt dataset for glacier inventories with high-resolution DEMs as a direct result. Results on very recent glacier changes are presented. Chapter 5 broadens the picture to a more regional one by comparing two other glacier inventories that have been established with the same or a similar methodology. In Chapter 6 we aim at increasing the temporal resolution between the existing inventories by modeling glacier changes using meteorological input parameters and verifying the results with the observed volume changes. This allows for consideration of temporally varying tuning parameters. Chapter 7 offers concluding remarks and an outlook on future work in this field. In Appendix A, the original papers and posters are reproduced and Appendix B finally presents a comparison of pictorial documents with today's reality in the glaciological context.

2. GLACIER AND CLIMATE CHANGE

This chapter intends to offer a review on glacier changes in the Alps. The author tries to find a compromise between getting too far into details and a too short overview. It is a necessary basis to put the order of magnitude of very recent glacier changes, which will build the core of this thesis, into the right spatial and temporal context.

2.1. BEFORE THE PLEISTOCENE

There is evidence for ice on the planet Earth as far back as the Precambrian at the so-called Huronian glaciation (approx. 2000-2500 Myrs b.p.; Evans and others (1997)). Since then, at least 5 phases in the Earth's climate history suggest favorable conditions for major glaciations such as in the late Precambrian or in the Paleozoic Era (Nesje and Dahl (2000); Bradley (1985)). These early glaciations occurred at times, when the supercontinent Gondwana occupied the south polar regions (Crowell, 1999). Figure 1 shows a general global temperature and precipitation curve relative to today's conditions. The cold periods of the Precambrian and Paleozoic were interrupted by long periods of significantly higher temperatures than today.

The Alps started to evolve between 135 and 35 Ma b.p. during the late Tertiary (Grotzinger and others, 2007) and it is the late Cenozoic (about 34 Ma b.p.) that marks the beginning of the current 'icehouse' and significant glaciation periods starting in eastern Antarctica (Ehlers and Gibbard, 2007). Since then, ice sheets and glaciers have started to grow and wane without ever totally disappearing again (Barker and others, 2007, Zachos and others, 2001). Glaciation in the Northern Hemisphere started significantly later (around 3.2 Ma b.p.; Zachos and others (2001)).

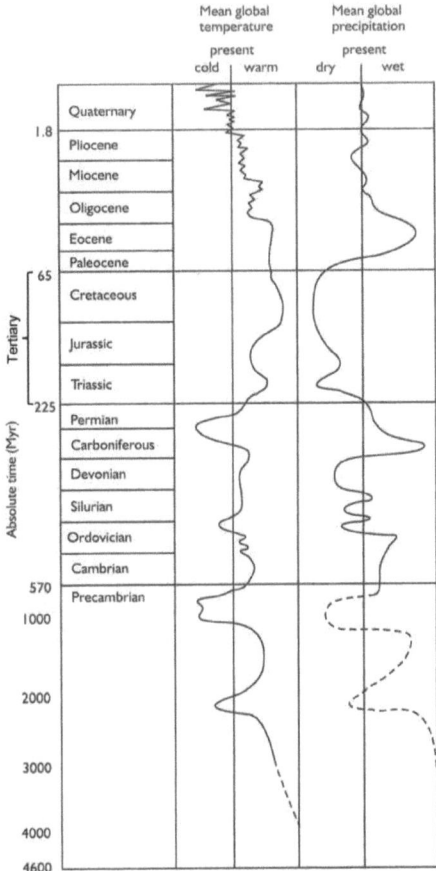

Figure 1: General global temperature and precipitation curves - note variable timescale. (adapted from Nesje and Dahl (2000)).

Figure 2 shows the occurrence of glaciation in Europe throughout the Cenozoic with the earliest records in Iceland and Norway. The onset of glaciation in the Alps was then comparably late, possibly not before MIS 22 (Marine Isotope Stage) which is at approximately 0.9 Ma b.p. (Ehlers and Gibbard, 2007, Muttoni and others, 2003).

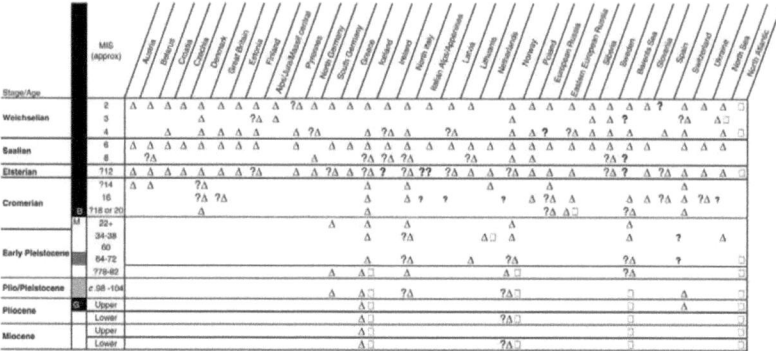

Figure 2: Occurrence of glaciation in Europe through the Cenozoic (Ehlers and Gibbard, 2007).

2.2. ICE AGES

In the Quaternary (approx. the last 2 million years up to and including the present time), the cycle of glacials and interglacials started to evolve and happened in a distinctive way since at least the last million years. Croll (1867a, b) first developed the theory that the repeating cycles of ice-ages could be explained by changes in the Earth's orbit and the axis' orientation. Milankovich later elaborated Croll's ideas and made them more popular (Milankovitch, 1941). The earth orbit's changing shape (eccentricity), the tilt of the earth's axis (obliquity) as well as it's wobble around its axis (precession) affect both the total energy input from the sun towards the Earth and the latitudinal energy distribution. Hays and others (1976) were able to verify Croll's and Milankovich's ideas about 35 years later by analyzing ocean sediment cores and the spectral analysis of $\delta^{18}O$-variations therein. The changes in astronomical parameters alone are the general 'pace-maker' of Quaternary climate change as widely accepted, however, it is obvious that many other complex factors influence it on different scales. Alterations in bio-geochemical cycles, tectonic and volcanic activity, the location of land-masses, oceanic circulation and various feed-back mechanisms should be mentioned (Nesje and Dahl, 2000).

During the past 2.5 Myrs, about 50 glacial/interglacial cycles occurred, whereas in the past 800 kyrs about 10 cycles are known. The cycles are well visible in a deep-sea oxygen isotope record from the equatorial Atlantic (Figure 3).

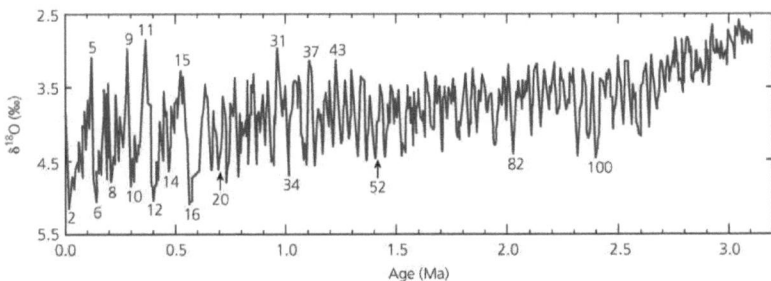

Figure 3: Deep-sea oxygen isotope record from the equatorial Atlantic. Selected marine isotope stages are labelled. (Fig. from Benn and Evans (2010) after Bradley (1999)).

Van Husen (2004) investigated the maximal extent of quarternary glaciations in the Eastern Alps and concluded that the penultimate of the quarternary's four major glaciations (called 'Rissian', MIS 6, approx. 130 ka b.p.) was larger than the 'Würmian' (MIS 2, approx. 21 ka b.p.) that is commonly referred to as the last glacial maximum (LGM).

2.3. THE LAST GLACIAL MAXIMUM

The world-wide ice-maximum of the last ice-age occurred between 24 and 19 ka b.p. (Ivy-Ochs and others, 2004). In the Eastern Alps glacier extent was limited to tributary valleys of the main trenches until about 24 ka b.p. despite a very cold climate for already 7 Millenia before. This is mainly explicable with topographical constraints (van Husen, 1997). Between 24 ka b.p. and 21 ka b.p. a rapid ice build-up lead to extended ice streams and to piedmont glaciers in the foreland. The build-up ended at ca. 21 ka b.p. and the 'Hochstand' lasted around 3000-4000 years (van Husen, 1997).

The maximal glacier extent of the Eastern Alps during the last glaciation is well documented (e.g. Ivy-Ochs and others (2008); Van Husen (1987, 1997, 2004)). Figure 4 shows the approximate extent of the Alpine glaciers at that time as dated by terminal moraines (Ivy-Ochs and others, 2008). The locations of many of the major lakes in the Alpine forelands are related to those days' terminal glacier lobes (e.g. Ammersee, Chiemsee, Lago di Garda, (van Husen, 2004)).

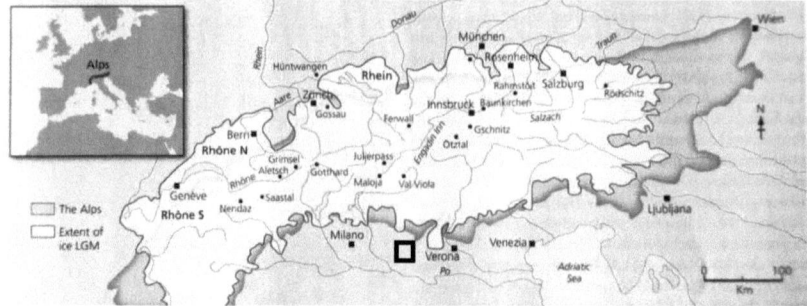

Figure 4: The glacier extent in the Alps at the end of the last iceage (Würm-stadial, LGM). The ice-sheet almost reached the location of today's Munich or Milan. Salzburg and Zurich for example were covered with ice (figure from Benn and Evans (2010) after Ivy-Ochs and others (2008)). The black rectangle shows the approximate extent of Figure 5.

Some of the terminal moraines of the LGM are still impressively visible as for instance South of today's Lago di Garda, Italy. Figure 5 shows a Google Earth image of this area and in the central lower part of the figure wavy moraine-structures are well visible.

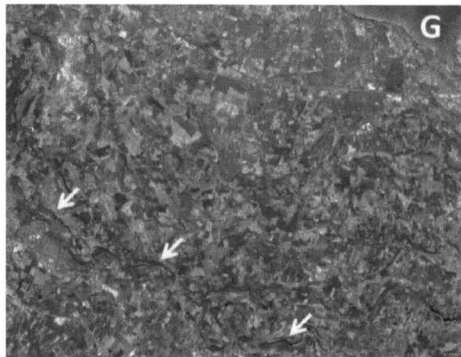

Figure 5: Terminal moraine-structures south of Lago di Garda, which is marked with G. The wavy structures near the white arrows are associated with the LGM terminal moraines (image from GoogleEarth). The approximate location of this figure's extent is displayed as a black rectangle in Figure 4.

In the Austrian Alps, the major valleys were ice-covered. Only the highest peaks exceeded the large ice-streams. Van Husen (1987) reconstructed extent and surface elevation of the Eastern Alps during the LGM. Figure 6 gives an example of a region in the central Eastern Alps. Above today's location of Innsbruck for instance a more than 1700 m thick ice-cover prevailed; the highest peaks were rocky Nunataks. According to these facts, it is very likely that the surface elevation of the highest Eastern Alpine firn areas have not changed dramatically since the LGM until today.

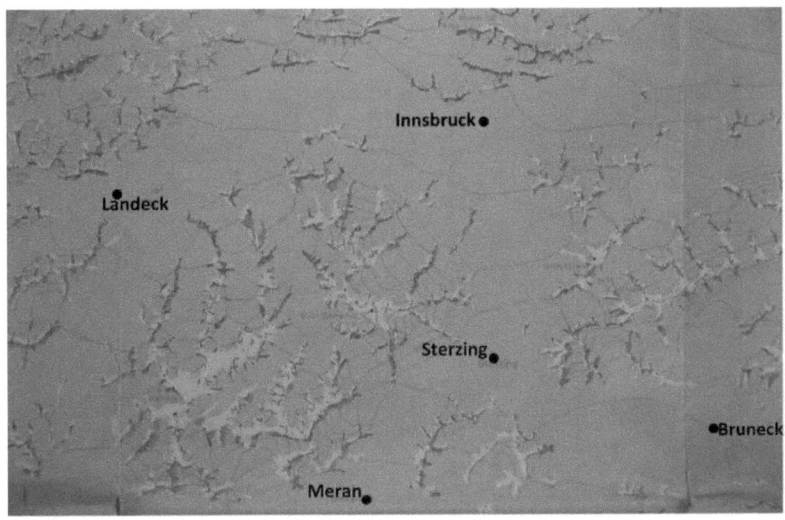

Figure 6: The ice-streams around the central Austrian Alps at the time of the LGM (Van Husen, 1987).

2.4. DEGLACIATION HISTORY

The early deglaciation after the LGM is believed to have occurred very quickly. Van Husen (1997) estimates that some hundred to a thousand years have been all that it took for the loss of about 50% of the glacier lengths. He conjectures that calving into lakes that had formed in overdeepenings below large LGM-glaciers might have been a reason for this quick recession.

Two short oscillations of active glaciers in contact with inactive ice occupying the main valley floors follow the quick recession after the LGM around 16 ka b.p. and have earlier been referred to as the Bühl- and Steinach-Stadial (e.g. Mayr and Heuberger (1968); van Husen (1997)). Nowadays, oscillations of that time-period are often referred to as the so-called Gschnitz-Stadial, named after a well-defined and well-investigated moraine in the Austrian Gschnitz-valley (Ivy-Ochs and others, 2008).

Following the Bölling/Alleröd interstadial, glaciers readvanced during the Younger Dryas cold periods significantly (Maisch, 1981, Patzelt, 1972). These glacier advances are commonly summarized as the Egesen-stadial (c.f. Heuberger (1966), Ivy-Ochs and others (2009), Maisch (1981); Figure 7). The end of the younger Dryas marks the transition from the late Pleistocene to the early Holocene which is commonly defined to have taken place at 11.7 kyrs b.p. (Nesje and Dahl, 2000).

Figure 7: An example of a distinct glacial trimline in the inner Ötz-valley. It is likely that this trim-line stems from the Egesen-stadial (photo: J. Abermann, July 2007).

2.5. THE HOLOCENE

Figure 8 shows the qualitative climate evolution during the Holocene. After the Egesen-Maxima, glaciers started to retreat with interruptions like the Kartell-stadial around 11 kyrs b.p. that was named after a moraine-complex in the very western-Austria (Fraedrich, 1975; Sailer, 2001; Kerschner and others, 2006). The very early Holocene was a period with increasingly dry but still comparably cold conditions and it was rather humidity that caused glaciers to retreat significantly at this time. Rock glacier systems developed in this period (Ivy-Ochs and others, 2009).

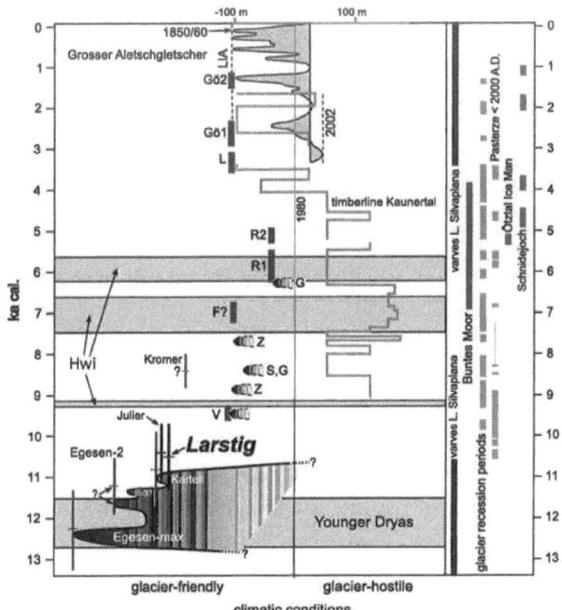

Figure 8: Qualitative summary of glacier variations and climate evolution during the Younger Dryas and the Holocene (Figure from Ivy-Ochs and others (2009)). Glacier advances are shown as grey areas. Three Holocene warm intervals (Hwi) are indicated with orange bars. The green line indicates variations in the Kaunertal timberline.

Around 10.5 kyrs b.p. a distinct shift towards generally warmer and likely also drier conditions took place that lasted almost uninterruptedly until about 3.3 kyrs b.p. A few interruptions such as the Venediger-oscillation prior to 9.2 kyrs b.p. (Nicolussi and Patzelt, 2001, Patzelt, 1972), the Kromer-stadial around 8.4 kyrs b.p. (Gross and others, 1978, Kerschner and others, 2006) or the so-called 8.2 kyr-event (Rohling and Palike, 2005) should be mentioned.

Figure 9 shows the image of a moraine at Rofenwiesen, Ötztal Alps (46°51'20.60"N, 10°52'47.75"E), marking a glacier extent well within the younger Dryas moraines but indicating a larger extent than during the little ice age (LIA) - it is suspected by the author that this is a remnant of one of these early Holocene glacier fluctuations. A detailed dating has not been done yet and could give evidence for that.

Figure 9: Moraine at Rofenwiesen, Ötztal (photo: J. Abermann, Sept 2009).

There is evidence that smaller glaciers have advanced between 6.3 and 5.0 ka b.p. also referred to as the so-called Rotmoos oscillation, whereas larger glaciers remained small (Bortenschlager, 1984, Patzelt, 1977). At 5.1 ka b.p. no ice existed on Tiesenjoch (3208m) in the Ötztal Alps as proven by the discovery of the Ötztal Ice Man (Bonani and others, 1994). Timberline was higher than today for most of the period as several findings of wood at many locations in the Eastern Alps (Nicolussi and others, 2005, Nicolussi and Patzelt, 2001) as well as in the Western Alps (Joerin and others, 2006) show. This obviously also implies the fact that glaciers were smaller than today several times during this period.

After 3.3 ka timberline moved down to lower altitudes with more frequent occurrences of glacier advances and shorter recession periods finally leading to the LIA-advances from the 14th century AD until 1850/60 (Grove, 2004a,b, Grove and Switsur, 1994, Holzhauser and others, 2005, Patzelt, 1973). Between 3.0 and 2.3 ka b.p. numerous glacier advances have been reported in literature, that are summarized as the so-called Göschner-1 oscillation (Deline and Orombelli, 2005, Patzelt and Bortenschlager, 1973, Zoller and others, 1966). Roman artefacts at Schniedejoch, Switzerland (Suter and others, 2005) as well as dendrochronological investigations (e.g. Nicolussi and others (2005)) indicate a period with little ice cover around 2.11 - 1.74 ka b.p. with glaciers probably smaller than around 1920 AD (Nicolussi and Patzelt, 2001, Schlüchter and Joerin, 2004).

In the early medieval a distinct glacier advance occurred in the Alps and is summarized as the Göschener-2-oscillation (around 600 - 800 AD; c.f. Orombelli and Mason (1997); Zoller (1960) with maximal extents almost as large as at their LIA-maximum (Nicolussi and Patzelt, 2001).

Large parts of the Medieval were characterized by a comparably small glacier extent and it is believed that glacier cover was then comparable to the one in the late 20th century (Nesje and Dahl, 2000).

2.6. THE LITTLE ICE AGE

The largest historical glacier fluctuations occurred during the LIA that is the period between the medieval warm period and the warm period starting in the first half of the 20th century (Grove, 2004a,b). It was first introduced into scientific literature by (Matthes, 1939) with extensive research later on that is comprehensively summarized in (Grove, 2004a,b). Climatically, the LIA was possibly triggered and primarily driven by a prolonged minimum in sunspot activity and solar output known as the Maunder Minimum between 1600 and 1700 AD and the Dalton Minimum around 1800 AD (e.g. Beer and others (2000)). Additional cooling appears to have been driven by stratospheric loading through volcanic aerosols such as through the eruption of Huaynaputina in 1600 AD (Thouret and others, 2002) or the Tambora in 1815 (Free and Robock, 1999). The definition of LIA in terms of a climatic vs. a glacier-historical concept including spatial distribution, onset and termination has recently been discussed critically by Matthews and Briffa (2005).

In many parts of the world glaciers expanded and fluctuated to larger extents than they occupied in the centuries before or after this generally cooler interval (Grove, 2004a,b, Matthews and Briffa, 2005). The synchronization of these advances varies widely and has been investigated on a smaller scale within Europe opposing Alpine to Scandinavian fluctuations (e.g. Nesje and Dahl (2003); Nussbaumer (2010)). Whereas the maximum extents in the Alps have been reached around 1600/1640 and 1820/1850 in the Alps, it was later in Norway (1750, 1870-1890).

Glacier fluctuations occurred in various steps and had their first maximal extents in the early 17th century. An extraordinary glacier length record based on historical documents, maps and measurements exists for Mer de Glace, France, for which at least 7 advances could be defined since 1600 (Nussbaumer and others (2007), see Figure 10).

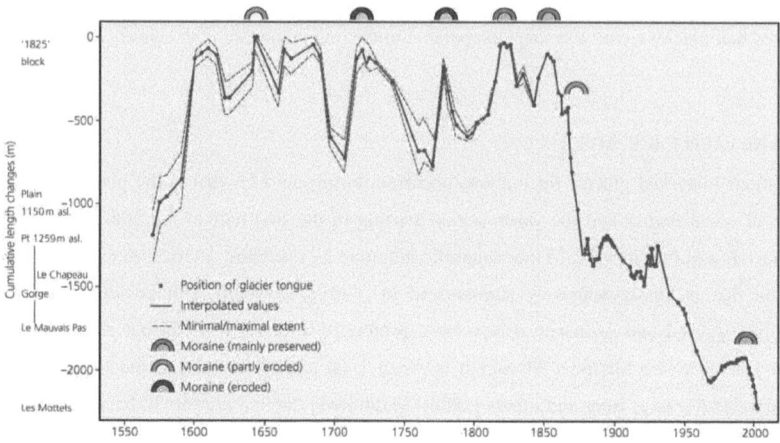

Figure 10: Fluctuations of Mer de Glace, France, during and after the LIA. The length-change curve is reconstructed from a variety of sources (Figure from Benn and Evans (2010) after Nussbaumer and others (2007)).

Similarly in the Eastern Alps, the advances of Vernagtferner have been documented well since as early as 1600 AD. Major advances that resulted in the damming of a glacier lake and the subsequent devastating drainage have been reported around 1600, 1679, 1773 and 1848 (Figure 11, Winkler (1996)). Until today, Vernagtferner is one of the best investigated glaciers in this area with mass-balance measurements (Escher-Vetter and others, 2009, Reinwarth and Escher-Vetter, 1999), hydrological measurements (Baker and others, 1982, Braun and others, 2000) and glacio-meteorological measurements (Weber, 2005).

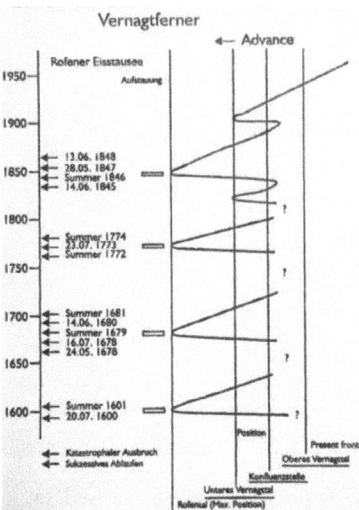

Figure 11: History of advance and retreating periods of Vernagtferner, Ötztal Alps including the information whether the advance had caused the damming of a glacial lake. From Nesje and Dahl after Winkler (1996).

A fascinating record from these early days of glaciological research stems from Walcher (1773), a professor and Jesuit priest, who was sent from Eastern Austria to Tyrol in the summer of 1772 in order to document the threatening events there (dammed lakes in Gurgler valley and Rofen valley, advancing Latschferner). He was a dedicated artist and in his charming book he attached several informative copper-plates of which one is reproduced in Figure 12 showing the Gurgler ice lake before it drained. He gave a detailed chronology of previous events of dammed lakes in the area and mentions his ideas how severe damage can be reduced (e.g. removing the ice with shoveling, drilling holes into the dam, firing cannons or trying to fill the lake by throwing rocks into it).

Figure 12: Gurgler Ferner (G, F), the ice-dammed lake Gurgler Eissee (A) and Langtaler Ferner (B). Copperplate published in Walcher (1773).

Figure 13: Lateral LIA-moraine of Marzellferner, Ötztal Alps (upper arrow) in 2010. The position of the lower arrow (2) indicates the location of a small re-advance, probably around 1920 (photo: J. Abermann, Sep 2008).

Generally, the maximal LIA glacier extent occurred between 1820 or 1850 depending on the glacier's geometry, its response time and the geographic location. Throughout the Alps, the most distinct moraine-features in today's glacier forefields can be attributed to the LIA-maximal extent, see Figure 13 for an example.

Gross (1987) compiled glacier outlines for the LIA maximum extent of the Austrian glaciers evaluating moraine features. He estimated the total glacier covered area in Austria to have been 1011 km² in 1850 (see Figure 15).

2.7. AFTER THE LITTLE ICE AGE UNTIL 1969

Another source of glacier area in the late 19th century is the 3rd military survey from 1870 that shows glacier extent shortly after the general maximum in great detail. Figure 14 gives an example of inner Ötztal. The glacier-dammed Gurgler lake was still there but moraine features (e.g. northern part of Marzell Ferner) indicate that glaciers have shrunk already since the maximum extent. Richter (1888) used these maps in order to create a first descriptive glacier inventory which gives us information on the glacier recession patterns of this period.

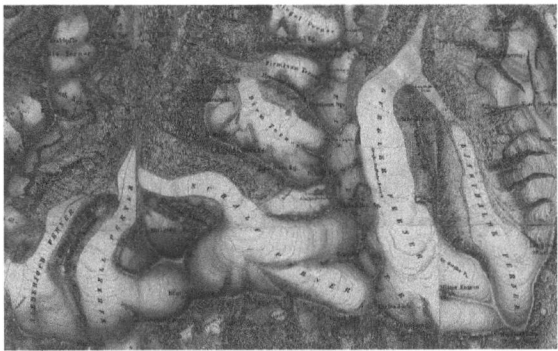

Figure 14: The 3rd military survey from 1870, example in the Inner Ötztal with e.g. Niederjoch-, Marzell-Gurgler- and Langtalerferner (Fig. from http://www.tirol.gv.at/tiris).

Gross (1987) estimated the first post-LIA-recession period to have lasted until 1895 with a minor interruption around 1870/1875. Around 1895 about 50% of the glaciers re-advanced and started to recede in the first decade of the 20th century. Following an Alps-wide advance around 1920, glaciers in 1925 had an extent similar to that in 1895 (Gross, 1987). In some of today's glacier forelands moraines from that time period are still visible - between the 1850 hochstand and today's glacier extent (e.g. Figure 13).

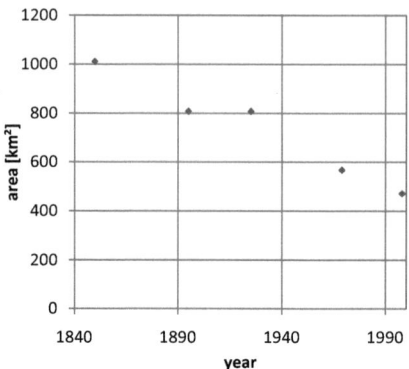

Figure 15: Evolution of Austria's post-LIA glacier-covered area. The numbers for the years 1850 and 1925 are derived from moraines. Length change measurements indicate no overall losses between 1895 and 1925. The values for 1969 and 1998 are the results of the two complete Austrian glacier inventories (c.f. Gross (1987), Lambrecht and Kuhn (2007), Kuhn and others (2009)).

Alpine glacier area and volume evolution since then started to be strongly influenced by the 20th century warming and enhanced solar radiation (e.g. Huss and others (2009). Especially the years after World War II show a pronounced warming trend (see Figure 16). Absolute rates of glacier area changes were comparable between 1925 and 1965 to the ones right after the LIA maximum - in relative terms they were stronger (see Figure 15).

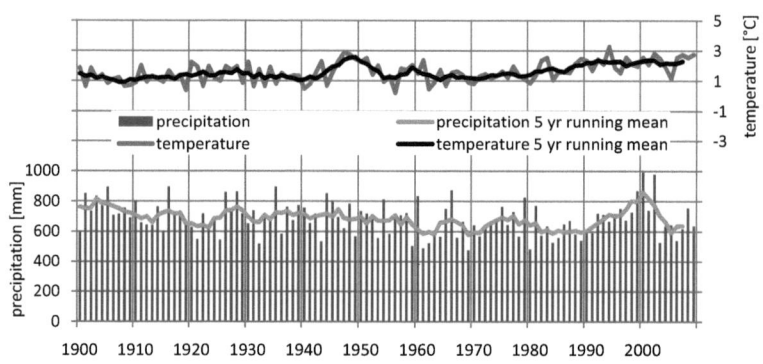

Figure 16: Evolution of annual mean temperature and precipitation in Vent, Ötztal, including the respective 5 years-running means. Data from the Institute of Meteorology and Geophysics; courtesy F. Pellet.

2.8. THE INVENTORIES 1969 AND 1998

The first complete glacier inventory for the Austrian glaciers was compiled in 1969 at the University of Innsbruck (Gross, 1987, Patzelt, 1978, 1980). Extraordinary favorable weather conditions and strong logistic endeavors (pers. comm. G. Gross) allowed for the coverage of all Austrian glaciers with aerial photographs within one week, which were then used for the photogrammetric production of glacier maps. These maps contain elevation contours, spot heights, glacier boundaries as well as snowlines on a scale of 1:10000 and 1:5000 (Lambrecht and Kuhn, 2007; Kuhn and others, 2009).

We are very fortunate to be in use of this inventory at that date for several reasons. First of all, compared to international standards, the efforts for a glacier inventory started visionarily early. It was the international hydrological decade (1965 - 1974), that increased scientific interest in glaciology significantly and lead to combined international efforts to start the world glacier inventory (WGI) (Ohmura, 2009, Zemp and others, 2007). The Austrian national inventory of 1969 is one of the first complete ones. As of 2008, still about 46% of the world's glacier area were not covered by neither the WGI nor the later-initiated GLIMS-dataset (Ohmura, 2009). Since then, this percentage has decreased through ongoing efforts (e.g. Paul and others (2009)).

Besides that, the late 1960s mark an interesting period in which the general pattern of glacier recession of the 20th century started to be interrupted by a significant glacier advance in the Alps (e.g. Patzelt (1985), for an example see Appendix A). Later in this thesis, an attempt will be made in order to reconstruct this advance and thus find the climatic reasons for it. These considerations are summarized in Abermann and others (2011b).

The period of glacier advances lasted approximately until 1985. Since then, glaciers continued their recession practically uninterrupted (and in the latter part accelerated; c.f. Abermann and others (2009b)) until today. The area, mass and volume gain after 1969 was quickly lost again and glacier changes were strong enough in order to start planning a new inventory. In total, rates of area losses were less negative than between the advance of 1920 and 1969 (see Figure 15). The survey flights for the new inventory were carried out by the Division of Aerial reconnaissance of the Austrian Army (Kuhn and others, 2009) in collaboration with the Federal Office of Metrology and Surveying (BEV), Vienna. Because of the requirement of cloud-free and snow-free scenes, it took several years until all the data was acquired (Lambrecht and Kuhn, 2007) and the aerial photographs thus span the years 1996 to 2002. However, most glaciers (73% by number and 81% by area) could be covered in the years 1997 and 1998. The image scale varies from 1:15000 to 1:35000 and aerial photographs exist partly in color, partly in black and white.

From the aerial photographs, a new approach for semi-automatic DEM generation was developed and applied. Details for the procession can be found in Eder and others (2000) or Würländer and

Eder (1998). The final results are high-quality DEMs with a mean vertical accuracy of +/-1.9 m on a 10m-raster that has later been interpolated to a 5m-raster.

The glacier delineation procedure was done manually on the base of a stereo model and orthoimages. The inventory of 1969 was digitized during the compilation of the new inventory and both boundaries have been imported into a GIS including additional information such as drainage area, ID-number, glacier name and area, aspect of the ablation and accumulation area. Combined with the DEMs, related products such as area-elevation distributions, minimum, maximum and mean elevation of the glaciers could be derived. In order to offer comparability for one fixed time period, Lambrecht and Kuhn (2007) present a temporal homogenization for 1998 based on a degree-day approach. In this way, the value for the extent of glaciers that have not been surveyed in 1998 (i. e. between 1996 and 2002) was calculated. However, the sum of all glacier extents from the different dates differs from the homogenized sum by 1.5% only and does not introduce any significant difference.

3. GLACIERS IN AUSTRIA

The inventory work that has been done and published so far allows for putting the results into a glaciological as well as a regionally climatic context. As the inventory of 1998 is the latest complete one, this chapter aims at using this fairly recent inventory as a basis for giving the reader a general idea about the current glacier distribution in Austria starting with an overview of glacier types in section 3.1., continuing with the relationship between basic quantities (such as aspect, minimum, median and maximum elevation, size) in section 3.2. that finally lead to the presentation of the first paper that is related to this thesis in section 3.3., wherein the climatic controls on the current glacier distribution and the recent glacier changes are investigated.

3.1. TYPES

There are various ways to classify glaciers and it seems generally informative to do that due to differences in their sensitivity or genesis. However, the author does not see the significance of a too detailed division and thus sticks to a very broad and general one exemplarily.

3.1.1. VALLEY GLACIERS

Most of the larger glaciers in Austria are valley glaciers. They are characterized by a comparatively big vertical extension that allows for transportation of high elevated accumulation down to low valleys. Their volume change currently contributes most to the total volume change compared to other glacier types. Figure 17a shows an example of a valley glacier in the Ötztal Alps, in this case Langtalferner with a vertical extension from 3400m down to 2500m a.s.l.

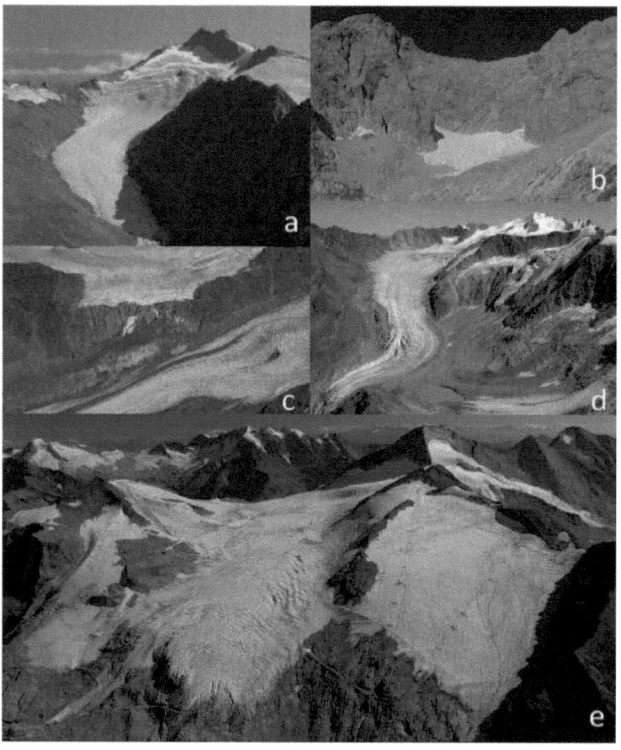

Figure 17: Glacier types in Austria: a) Valley glacier, e.g. Langtaler Ferner, b) cirque glacier, e.g. Mieminger Schneeferner, c) regenerated or rejuvenated glacier, e.g. Gepatschferner western margin over Langtauferer Ferner, d)hanging glacier, e.g. upper part of Taschachferner and e) a 'plateau' glacier, e.g. Gefrorene Wand Kees (photos: J. Abermann, Sep 2008).

3.1.2. CIRQUE GLACIERS

There are many cirque glaciers in the Eastern Alps and they can be divided into two groups according to their genesis and thus sensitivity. On the one hand they can be remnants of larger ice masses that shrank to a glacier extent as a response to climate change. In this case they are very sensitive to climatic change in that they may rapidly disappear and reform (Kuhn, 1995). The second group however, owes its existence to favorable topographic conditions. Accumulation in the cirques occurs mainly through avalanches that fill up to a rather constant extent and volume every winter. This kind of cirque glaciers is much less sensitive to temperature changes but more sensitive to changes in precipitation and thus reacts differently to an imposed climate change than valley glaciers do (Kuhn, 1995). In the case of Mieminger Schneeferner for example (Figure 17b), ice extent has not been significantly larger during the LIA due to topographic constraints and an early observation does not indicate large changes compared to today (Kilger, 1892, Kuhn, 1993).

3.1.3. REGENERATED GLACIERS

A regenerated glacier that is sometimes also called rejuvenated glacier is a glacier which develops from ice avalanche material beneath a rock cliff. Very few glaciers of this type are found in Austria, an example is given in Figure 17c where the western edge of the Gepatschferner plateau drops steeply into Langtauferertal where it merges with Langtaufererferner.

3.1.4. HANGING GLACIERS

Hanging glaciers do not connect to the main valley below and 'hang' in side valleys. An example could be the western part of Taschachferner in the upper middle part of Figure 17d.

3.1.5. PLATEAU-LIKE GLACIERS

In contrast to valley glaciers and cirque glaciers, plateau glaciers evolve when topography allows the formation and coverage of a generally flat mountain tract. Mass is transported away from the plateau in hanging glaciers down their edges. There are not many plateau-glaciers in Austria, Figure 17e shows the example of Gefrorene Wand Kees, probably the example that gets closest to this type of glacier.

3.2. GLACIER DISTRIBUTION, GLACIER CHANGES AND TOPOGRAPHY

In 1998, there were 915 glaciers in Austria, covering 470 km^2(Lambrecht and Kuhn, 2007). The largest glacier covered mountain groups are the Ötztal, Venediger and Glockner group, together covering more than 60% of the total glacier area in Austria (see Figure 18 for location, Table 1 for acquisition dates, area and number and Figure 19 for the fraction of total glacier area in the respective group). More information on the distribution connected with the climatic setting will be discussed in section 3.3. which is summarized in Abermann and others (2011b).

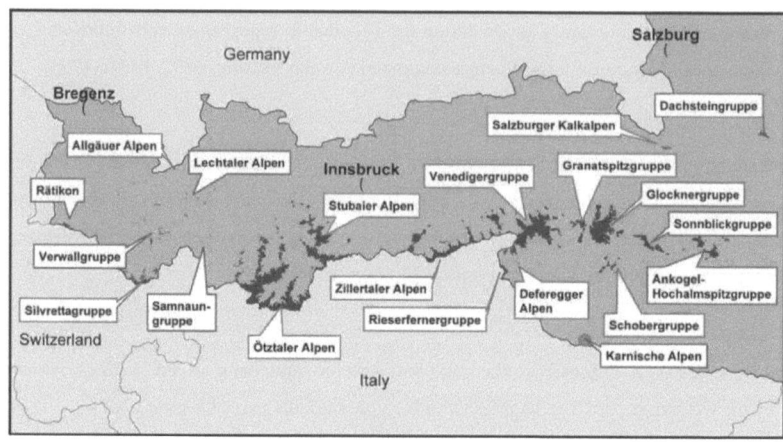

Figure 18: Overview of the glacier cover in Austria and the mountain groups as referred to in Table 1 (Figure from Lambrecht and Kuhn (2007)).

Table 1: Acquisition year of the two complete glacier inventories according to the mountain groups, area of the inventory in 1998, and number of glaciers in each mountain group sorted by the group's size.

mountain group	year inventory I	year inventory II	area 1998	# of glaciers
Ötztal	1969	1997	155.53	211
Venediger	1969	1998	81.01	101
Glockner	1969	1998	59.84	78
Stubai	1969	1997	53.00	117
Zillertal	1969	1999	51.72	136
Silvretta	1969	2002	20.33	48
Ankogel	1969	1998	16.15	52
Sonnblick	1969	1998	9.74	31
Granatspitz	1969	1998	7.52	31
Dachstein	1969	2002	5.83	8
Verwall		2002	5.08	35
Schober		1998	3.49	26
Rieserferner		1998	3.13	10
Hochkönig	1969	2002	1.87	3
Rätikon	1969	1996	1.60	2
Lechtal		1996	0.69	14
Defereggen		1998	0.43	7
Karnische			0.18	1
Samnaun		2002	0.11	3
Allgäu		2000	0.09	1

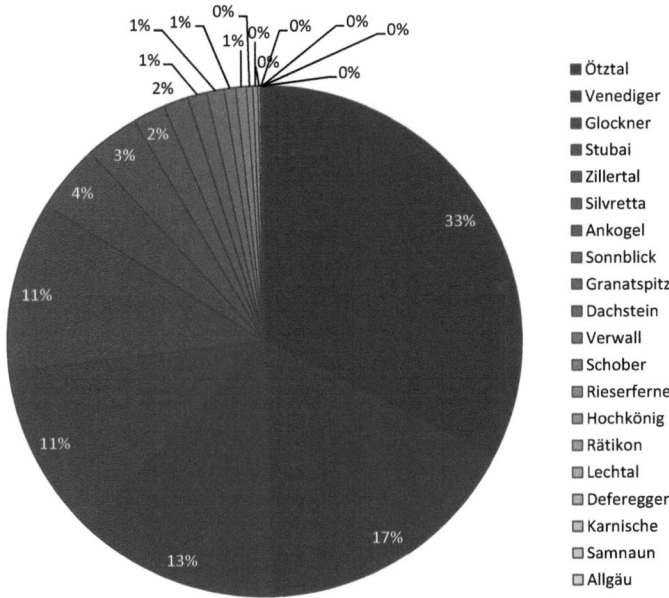

Figure 19: Relative contributions of individual mountain groups to the total Austrian ice cover.

Figure 20a-c show basic parameters such as size (Figure 20a), number (Figure 20b), maximum, median and minimum elevation (Figure 20c), and their relation to the glacier's aspect. Figure 20a and b show that by far the largest glacier covered area is directed towards NW over N to NE. This has two main reasons: At the latitude of Austria, areas directed towards north are more shaded and receive less global radiation. However, it is important to note that this picture would change slightly if the observed inventory was not restricted to the Austrian national boundaries. There are some glaciers on the south side of Silvretta, Ötztal, Stubai and Zillertal Alps that are covered in the Swiss (Kääb and others, 2002) and the South-Tyrolean glacier inventories (Knoll and others, 2009). The majority of Eastern Alpine ice cover around the main Alpine crest is however found in the Austrian part.

Figure 20c shows mean maximum, median and minimum elevation of the glaciers exposed towards the respective direction. It is interesting that maximum and minimum elevation does not vary strongly among the exposition classes. Median elevation shows an expected distribution: northerly-exposed glacier areas and median elevations are significantly lower (by about 150m on average) than southerly exposed glacier areas.

Figure 20d-f illustrates the dependence of the observed changes between 1969 and 1998 on the aspect of the glacier area.

Absolute volume change is displayed in Figure 20d. It is not surprising that the shape resembles the area distribution very much. It is thus more sensible to use the mean thickness change (Figure 20f) for comparing differences in volume change in a 'normalized' manner.

Relative area changes were strongest in the western-exposed section with a loss of more than 20% of the initial area (1969, Figure 20e). Smallest relative area losses were observed in the south-west exposed section with only about -12%. The glaciers with a component directed towards north show a rather homogeneous relative area loss of about -17% which is representative for the overall area changes observed.

Mean thickness changes are finally displayed in Figure 20f. An average value is a mean thickness loss of 8-10 m between 1969 and 1998 with a minimum in the east-exposed section (-7.5m).

The remarkable spike that is visible in Figure 20a, d and f in the southeastern corner with larger area (Figure 20a), larger volume loss (Figure 20d) and a remarkably larger mean thickness change (Figure 20f) compared to other areas directed towards south (i.e. S, SW), is related to Pasterzenkees, Austria's largest glacier. This valley glacier's comparably thick glacier tongue shows a substantial thickness loss in recent years (e.g. Kuhn and others (2009); p. 107) explaining the spike in absolute volume loss (Figure 20d) and a mean thickness change. It is also interesting to mention that the pattern of relative area change (Figure 20e) does not reflect this asymmetry corroborating the fact that volume and area change and their rates of changes evolve differently. This will be subject to more detailed discussion in chapter 4.

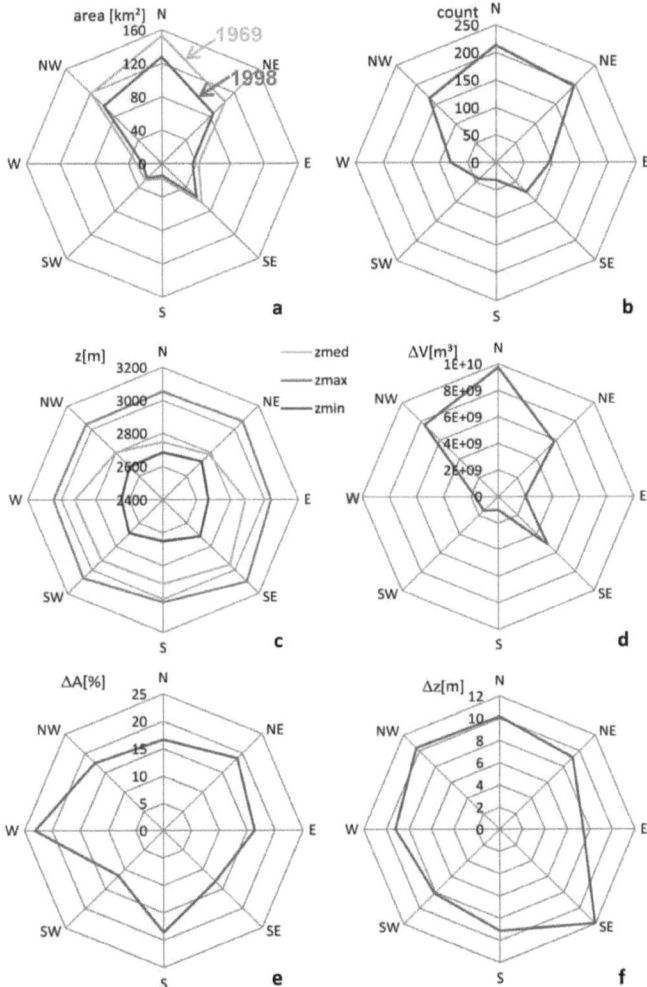

Figure 20: Glacier characteristics as a function of exposition: a) area 1969 and 1998, b) number of glaciers 1998, c) mean minimum, maximum and median elevation, d) volume loss, e) relative area loss and f) mean thickness loss.

With the following 3 figures (Figure 21a, b and c) a brief investigation of the relation of the same basic quantities as above with the glacier size is displayed.

Figure 21a shows that the majority of the glaciers in Austria in terms of total area lies in the class between 1 and 5 km² (i.e. more than 200 km² in 1969, about 195 km² in 1998). Most glaciers by number are in the smallest class (almost 400 glaciers are smaller than 0.1 km²). The fact, that these small glaciers cover a significant fraction of the total glacier area in Austria is important regarding the methodological decision of the most recent glacier inventory that will be introduced in chapter 4 (Abermann and others, 2010a, Abermann and others, 2009b). Very few glaciers are larger than 5 km² (less than 10) but cover in total more than a quarter of the total glacier area.

Figure 21b shows the dependence of mean minimum, maximum and median elevation of glaciers on the size class. The scatter of each quantity reduces with size as it is shown exemplarily in Abermann and others (2011b). However, mean maximum elevations rise with size, minimum elevations do the contrary. Large glaciers are in areas with high mountains, thus high maximum elevations. Accumulation areas can reach large dimensions and dynamics are the cause for the low minimum elevations. Glaciers between 5 and 10 km² show highest median elevations (on average just above 3000m).

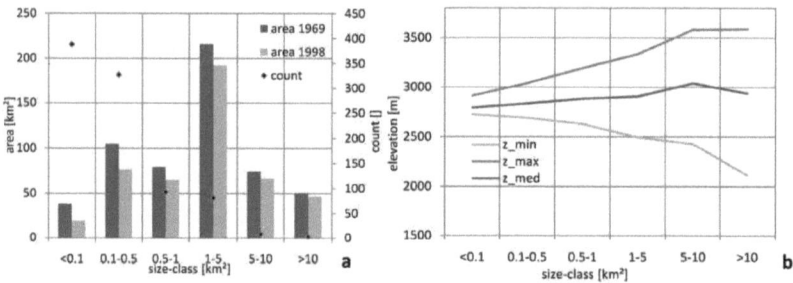

Figure 21: a) glacier area from 1969 and 1998 as a function of size class (left axis, bars); number of glaciers in a size class (right axis, points). b) Mean minimum, maximum and median elevation as a function of the size class.

Analogous to the consideration with aspect before, the dependence on glacier changes between 1969 and 1998 on glacier size is presented in Figure 22. The smallest class shows the strongest relative area changes (almost -50% in total), whereas glaciers larger than 5 km² only lost between 5 and 10%. For the mean thickness change it is the other way round: larger glaciers lost significantly more (up to -14m), whereas hardly any differences are visible for glaciers smaller than 1 km². Geometry is the key to understand this discrepancy: Large thick glacier tongues show rapid downwasting which does not necessarily results in a strong area change due to the valley geometry.

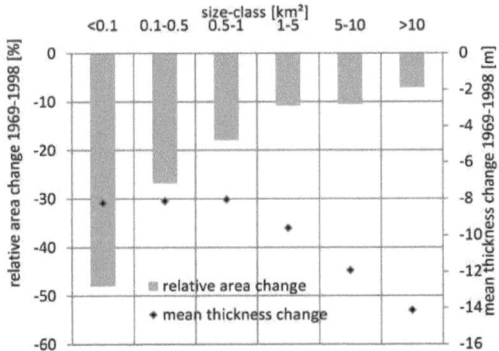

Figure 22: Relative area change between 1969 and 1998 referring to the area of 1969 (left axis, bars), mean thickness change (right axis, points).

3.3. GLACIER DISTRIBUTION, GLACIER CHANGES AND CLIMATE

The topographic constraints and their relation to glacier cover were introduced in section 3.2. This section refers to Abermann and others (2011b), which is reproduced in Appendix A1. It investigates the connection between glacier distribution and the general climatic setting. Furthermore, a qualitative assessment of the main meteorological driving factors in the Alps (summer temperature, winter precipitation, c.f. Kuhn and others (1999)) and their temporal evolution is presented.

4. VERY RECENT GLACIER CHANGES

After the compilation of the second complete Austrian glacier inventory, the trend of negative area and mass evolution accelerated. The mass balance season of 2002/2003 revealed a negative record on many long-term studies (Fischer, 2010) which is all consistent with the generally intensely studied climatic change (Solomon and others, 2007).

The long-term studies of length changes as well as mass balance indicated that the rates of glacier losses have increased and it became increasingly notable that the originally conceived interval for re-inventorying of 50 years (Ohmura, 2009) does not meet the requirements to sensibly monitor the rapid glacier changes of the past decades.

A multi-disciplinary project to investigate the potential and future perspective of glaciological monitoring strategies under changing climatic conditions was carried out between 2001 and 2004 (OMEGA-project) in which a number of airborne laser scanning (ALS) surveys of different glacier sites were made. This project resulted in a wealth of data that is summarized in Kuhn (2007) and Pellikka and Rees (2009).

In 2006, the regional government of Tyrol started with an airborne laser scanning (ALS) campaign covering the whole federal state. The data from the OMEGA-project together with the data from the regional government provided the base to summarize the next two papers of which the first gives a methodological overview on the potential of high-resolution DEMs for glacier monitoring, with a special attention towards glacier inventory studies (Abermann and others, 2010a; Appendix A2). The second one is dedicated to recent glacier changes in Austria's largest glacier covered area, the Ötztal Alps and on the quantification of recent rates of changes (Abermann and others, 2009b; Appendix A3).

5. REGIONAL CONSIDERATIONS

A remarkable coincidence is that a similar dataset with the same acquisition years of the two recent Austrian glacier inventories was established for a neighboring area in South-Tyrol, northern Italy by (Knoll, 2006, Knoll and others, 2009). Furthermore, the Austrian inventory of 2006 has so far been extended spatially to all areas where the data is already available (Stubai Alps: Seiser (2010); West-Tyrol: Goller (2010)).

A comparison of the results presented in chapter 4 with South-Tyrol and the Stubai Alps each resulted in a poster contribution at 2 international conferences. The posters are reproduced here and show significant differences of glacier changes. It is evident, that less mass and area loss between 1997 and 2006 occurred in the Stubai Alps (Abermann and others, 2010b), getting more negative in the Ötztal Alps and being most negative south of the main Alpine crest (Abermann and others, 2009a). Both posters are reproduced in Appendix A4 and A5.

6. GLACIER INVENTORIES: A CLIMATIC INTERPRETATION

This chapter aims at putting the observations that are unique in resolution and accuracy into a climatic context. Area and volume changes are known well between 1969 and 1998, and consequently for large parts until 2006. It was the challenge to use this detailed spatial resolution and improve our understanding of a regionally different temporal resolution by using meteorological data. The potential and limitations of the approach chosen can be assessed by comparing the results with a wealth of mass balance data.

Section 6.1. thus introduces briefly the different methods to obtain direct glacier mass balance data and section 6.2. the state-of-the-art concerning mass balance modelling. In section 6.3. we refer to a recent paper where we apply a positive degree-day model to the glaciers in Austria to improve the temporal course of the results (Abermann and others, 2011a; Appendix A6).

6.1. INTRODUCTION TO MASS BALANCE DETERMINATION

The surface mass balance (b_s) of a glacier directly reflects a cumulative signal of all meteorological processes impacting a glacier's surface. The range of physical processes involved is complex and can be summarized as

$$b_s = a_s + a_a + a_r + a_w - m - s - c \qquad (1)$$

representing snowfall (a_s), avalanche deposition (a_a), refreezing of liquid water (a_r), wind deposition (a_w), snow and ice melt (m_s), sublimation (s) and calving (c) (Cuffey and Paterson, 2010). A glacier in a steady state is in balance between the ablation and accumulation terms over a mass balance year which makes b_s equal to zero. In the Alps, many of these terms play a small to negligible role which makes simplified assumptions in many cases justifiable.

In praxis, there are three major ways to determine the mass balance of a mountain glacier, the direct glaciological, the geodetic and the hydrologic method. A fourth method should be mentioned for completion and because of its significance, although it is not applicable to glaciers of the scale studied in this thesis: Great advances have been made recently by measuring changes in the Earth's gravity field due to a changing distribution of ice masses from satellites. There are various publications, describing results of these recent developments concerning the ice sheets in Greenland and Antarctica, e.g van den Broeke and others (2009) or Velicogna (2009).

6.1.1. DIRECT GLACIOLOGICAL METHOD

The direct glaciological method is the most laborious method for deriving a glacier's mass balance. Direct measurements have been made for around 250 of the world's glaciers but less than 90 records extend over 10 years or more (Hoinkes, 1970; Braithwaite, 2002; Dyurgerov and Meier, 2005). A detailed overview on the practical aspects of measuring mass balance can be found in (Kaser and others, 2003a). Generally, the idea is to measure accumulation and ablation totals. For accumulation measurements, pits are excavated and a density profile of the snowpack is determined. Snow depth and density allow for calculation of the snowpack's snow-water equivalent. This is generally done at the estimated maximum of the annual cycle and in the Alps, May 1st is the established date for that. Ablation in turn, is measured by drilling stakes into the ice and regular readings of the relative lowering of the ice surface. A network of stakes and pits is needed in order to get beyond point measurements and to estimate a whole glacier's mass balance. Four long-term time-series of mass balance measurements are used in this thesis to validate a modelling approach as presented in section 6.3.

6.1.2. GEODETIC METHOD

The increasing availability of DEMs with improving quality makes the so-called geodetic method a very important option. Basically, DEMs from different points in time allow for calculating elevation differences (volume change). Compared to direct measurements, the advantage is that a whole glacier's volume change can be determined in contrast to interpolation from point measurements and also complex terrain can be covered equally well. Furthermore, the laborious and time-consuming field-work of the direct method is not necessary. A detailed overview on the state-of-the-art of the geodetic method for mass balance determination can be found in Bamber and Rivera (2007). There are several limitations on the comparability of directly measured and geodetically derived mass balances that are due to interpolation of point measurements, density uncertainties and glacier dynamics (vertical velocities). In this thesis, the geodetic method of mass balance determination exploiting two successive photogrammetrically derived DEMs is used in section 6.3.

6.1.3. HYDROLOGIC METHOD

The hydrological method of mass balance determination is based on the idea of using the knowledge or assumptions of other hydrological terms and determine glacier mass changes by attributing them to the residual (e.g. Collins (1984); Singh and Singh (2001)). According to Benn and Evans (2010):

$$P - R - E + \Delta S_g + \Delta S_s = 0 \qquad (2)$$

wherein P is precipitation, R is runoff, E is evaporation ΔS_g is the change in glacier storage and ΔS_s is the change in other water storage (such as non-glacial snow-cover, supraglacial, englacial or subglacial water, etc.). The spatially distributed knowledge of precipitation is one of the crucial parameters that very often is not given and has to be estimated and interpolated from point measurements. These point measurements (i.e. rain gauges) are subject to substantial measurement errors due to influence of wind or snow (c.f. Auer and others (2005); Yang and others (1998)). Runoff can be measured at gauges comparably well and evaporation is often estimated because measurement is difficult, especially at high altitudes (Barry, 1992). To discern ΔS_s from ΔS_g, hydrological methods are often used in conjunction with ablation measurements and modelling studies (Benn and Evans, 2010) and have proven to produce reliable results in different ice-covered areas (Kaser and others, 2003b, Kuhn, 2003). Under present climatic conditions, ΔS_s is way larger than ΔS_g, which is the reason why the latter is neglected in various studies (Kuhn, 2003).

6.2. INTRODUCTION TO MASS BALANCE MODELLING

From section 6.1. it is evident, that the derivation of mass balance is a challenge and crucial if the complex connection between the cryosphere and the atmosphere is attempted to be understood. There have been numerous efforts that date way back in the early days of glaciological research to create time series of a glacier's mass evolution from indirect measurements (e.g. temperature and precipitation). Methods vary substantially in complexity, depending on the scope of research, and, especially on the size of the study area. Ablation models can largely divided into two groups with various stages of refinements within these groups: Temperature-index models (section 6.2.1.) and energy balance models (section 6.2.2.). They have both in common, that they must be combined with accumulation models. In many studies, accumulation is estimated by interpolated precipitation datasets combined with a snow/rain-threshold (e.g. Kuhn (2003); Schaefli and others (2005)). Recently, advances have been made to achieve spatially more complex precipitation patterns, incorporating wind deposition or avalanche redistribution (e.g. Lehning and others (2008)).

6.2.1. TEMPERATURE-INDEX

Temperature-index models are based on the relationship between air temperature and snow or ice-melt in order to model ablation. This reliable connection has been first formulated for Suldenferner by Finsterwalder and Schunk (1887) and successfully applied on many locations in the world (e.g. Braithwaite and Zhang (2000); Hock (2003); Huss and others (2008); Kuhn (2003); Lang and Braun (1990)). The basic formulation of this approach is e.g. after (Hock, 2003):

$$b_{abl} = \begin{cases} (DDF) * T & | \ T > 0°C \\ 0 & | \ T < 0°C \end{cases} \qquad (3)$$

wherein b_{abl} is the ablation at a certain point in [m w.e.], T is the temperature in °C and DDF the so-called degree-day factor in m/(°C*d), which is generally assumed to depend on the surface properties, being larger for ice than for snow (Hock, 2003). The good performance of simple degree-day approaches is attributed to the fact that many components of the energy-balance (see section 6.2.2.) are strongly correlated with and thus implicitly captured by the temperature fairly well (Ohmura, 2001). Many studies include a potential short-wave radiation term to enhance the performance of degree-day-models at complex terrains to account for topographical shading (e.g. Hock and others (2007); Huss and Bauder (2009); Pellicciotti and others (2005)).

6.2.2. ENERGY BALANCE

In contrast to temperature-index models, a more complex and sophisticated approach to calculate ablation is solving the energy balance equation by determining the energy fluxes between the atmosphere and the snow or ice surface and the ones within the snowpack. In total, the energy fluxes have to equal zero, as no energy can be created or destroyed, and can be summarized after Benn and Evans (2010) as:

$SW + LW + QH + QE + QR - QT - M = 0,$ (4)

wherein SW is the net short-wave radiation flux, LW is the net long-wave radiation flux, QH is the sensible heat transfer, QE is the latent heat transfer, QR is the energy from rain, QT is the energy change used for temperature change in the ice, and M is the energy used to melt ice or freeze water. These models have been successfully applied at many locations in the world (Hock and others, 2007, Klok and Oerlemans, 2002, Machguth and others, 2009, Mölg and Hardy, 2004).

It is obvious, that the data requirements are much larger for energy-balance models (eq. 4) than for temperature-index models (eq. 3). Hock and others (2007) found in their detailed comparison of model performance, that there is no distinct answer to which approach is more successful. This has partly to do with the different quality of the input data (i.e. temperature is usually of higher quality than radiation components are) and partly with parameterizations that are necessary, and introducing errors with the energy balance approach. In order to resolve processes and their change with time, energy balance models can provide more insight.

6.3. CALIBRATING MASS BALANCE WITH GI-DERIVED VOLUME CHANGES

A logical continuation of the work presented so far, is the attempt to combine the spatially well resolved glacier inventory data with meteorological data in order to increase the temporal resolution of glacier changes. This has been performed in a recent publication for 96% of the glacier covered area in Austria. Data, methods, results, potential and limitations are discussed in Abermann and others (2010a), which is reproduced in Appendix A6.

7. CONCLUSIONS AND OUTLOOK

The submitted thesis starts with a general review on glacier changes in the Alps and after briefly examining the evolution in the world's cryosphere up until the LGM, the Eastern Alps are considered more in detail. The core of the research is the ice cover in Austria especially in the past 40 years for which very detailed data is available in the study area. A new dataset was applied in order to present a new way to determine the glacier extent in a detailed manner. Automatic methods are not applied in order to keep the quality of the data. A test area in the well-studied Ötztal Alps was chosen to show the applicability of such a new method. Rates of area and volume changes have been compared between the inventories in order to quantify the accelerated glacier changes in the region. Volume changes accelerated more than area changes did. It has been mainly attributed to topographic constraints. The regional glacier distribution was put into context with the general climatic setting; regional gradients of glacier changes are shown. It was found, that across the main Alpine divide, southern areas lost more glacier area and volume than in the North. An attempt was made in order to connect glacier changes for a large sample of glaciers in Austria. Specific net balances of glaciers where this information is available were reproduced with a correlation coefficient of 0.81. An outcome of this study is that modeling long-term glacier changes is very questionable with constant tuning parameters. By adapting the tuning parameters' seasonal and long-term course an improved fit between measured and modelled balances has been achieved.

In which way can this thesis be the framework and starting point for future studies?

There are several possibilities how the findings can be used and extended in future. A primary goal is the completion of the LIDAR-derived most recent glacier inventory. There are ongoing efforts, partly finished, to apply the method that has been presented in Abermann and others (2010a) to other regions in the Austrian Alps (Goller, 2010, Seiser, 2010). The remaining data is supposed to arrive soon and will make a third complete high-resolution inventory available. Unfortunately, the time-span of the LIDAR-DEMs that are and will be the basis for this inventory are not all from the same year but will span from 2006-2011 with the largest parts acquired in 2006. It may be necessary to think of a homogenization method to make glacier changes comparable. Lambrecht and Kuhn (2007) showed a way how this can be made relatively simply.

This will then be the basis to compare acceleration trends on a regional scale analogously as presented in this thesis but on a larger scale. Furthermore, it builds the basis for all modeling approaches of the recent past and also future. At this point, two different ways should be followed from the point of view of the author:

First of all, it would be easy and highly interesting how DDFs evolved in the recent past: Has the rise of DDFs that has been found in this thesis continued in very recent years? The same methods as applied in Abermann and others (2011a) together with the new inventory of 2006 and mass balance data will answer this question. Another dataset as meteorological input has to be thought of, because ERA40 ends in mid-2002. ERA-Interim could be a possible alternative, although it has to be investigated whether the temporal resolution of daily values is enough. Otherwise, the NCEP/NCAR reanalysis product could be an alternative (Kalnay and others, 1996).

Secondly, the modeling part could be refined in many ways. We showed that a reasonable performance can be achieved with a simple model. To resolve physical properties and attribute recent glacier changes quantitatively to components of the energy balance, a more physical approach may provide further insights. Increased computational capacity and an increasing availability of data (with improved quality) will help.

The main goal of this thesis was achieved: The regional understanding of recent glacier changes was improved and temporally higher resolved. Furthermore, very recent glacier recession was quantified with state-of-the-art methods. May others enjoy the read and take it as a basis for future studies!

APPENDIX A: PUBLICATIONS RELEVANT TO THE THESIS

A1: PAPER I: CLIMATIC CONTROLS OF GLACIER DISTRIBUTION AND GLACIER CHANGES IN AUSTRIA

By: Abermann, J., Kuhn, M. and Fischer, A. (2011). *Published in Annals of Glaciology, 59*(50), 83-90.

Climatic controls of glacier distribution and glacier changes in Austria

J. ABERMANN,[1,2] M. KUHN,[1,2] A. FISCHER[2]

[1]*Commission for Geophysical Research, Austrian Academy of Sciences, Dr. Ignaz-Siepel Platz 2, A-1010 Vienna, Austria*
E-mail: jakob.abermann@uibk.ac.at
[2]*Institute of Meteorology and Geophysics, University of Innsbruck, Innrain 52, A-6020 Innsbruck, Austria*

ABSTRACT. In this study we aim to connect glacier extent in 1998 with general climatic conditions, and glacier changes between 1969 and 1998 with climate change in the Austrian Alps. The investigations are based on two complete glacier inventories, a homogenized gridded precipitation dataset and European Centre for Medium-Range Weather Forecasts re-analysis (ERA-40) data of air temperature at different levels. A relationship between median glacier elevation, minimal elevation, the general elevation of the surrounding mountains and mean climatic values was found. In the Austrian Alps, the existence of glaciers at low elevations can only be maintained with above-average accumulation or strong dynamic ice supply. For debris-free glaciers, we found a limit of ~2080 m a.s.l., where mean summer temperatures (June–August) exceed 4°C. Glacier changes from 1969 to 1998 are strongly negative both in relative area and in mean thickness. There is a weak and regionally varying negative trend in precipitation over this period. A spatially consistent sequence of positive temperature anomalies in the early 1980s and after 1990 offers an explanation for the retreat. The study shows that the observed spatial variability of glacier changes is connected more strongly to the topographic differences than to a regionally different climate change signal.

INTRODUCTION

The extent and location of mountain glaciers depend on both topography and climatic conditions. These two factors are interdependent because of the altitudinal gradients in temperature and precipitation but also as a result of synoptic processes (e.g. prevailing weather patterns, orographic uplifting) and specific local factors (e.g. local snow distribution, avalanching, shading). Furthermore, glaciers transport ice from high elevations to the terminus such that in steady-state conditions, low-elevation melt equals dynamic resupply. Thus, conditions that affect the upper end of a glacier (i.e. above-average precipitation) also help determine the altitude of its lower end (i.e. below-average elevation of the glacier tongue). The combination of regional glacier inventories taken at different points in time with climate data provides an opportunity to study the connection between glacier distribution, glacier change and climate.

Attempts to understand this complex combination of factors have been made by Shea and others (2004) for the Canadian Rockies and by Schiefer and Menounos (2010) for British Columbia, Canada. The relationship between exposure to solar radiation and glacier distribution was investigated by Evans and Cox (2005) for glaciers worldwide, in greater detail by Evans (2006) for alpine glaciers, and by Evans and Cox (2010) for arctic glaciers only. Haeberli and Hoelzle (1995) relate inventory data to basic glaciological characteristics in the European Alps; Hoelzle and others (2007) derive estimates of surface mass balances from glacier inventories and a parameterization scheme comparing the European Alps with the Southern Alps of New Zealand. Recently, the glacier setting in northern Italy (Knoll and others, 2009) has been described comprehensively and Bolch and others (2010) investigate regionally different glacier changes in western Canada based on Landsat scenes. This dataset is also used in various publications on glacier changes in the Swiss Alps (e.g. Kääb and others, 2002).

In this study the Austrian glacier inventories (Patzelt, 1980; Gross, 1987; Lambrecht and Kuhn, 2007; Kuhn and others, 2009a), which are unique in both accuracy and timespan, are used to investigate the relationship between: (1) glacier location and representative climatic parameters; and (2) glacier changes as a direct result of climate change.

STUDY AREA

The investigations are performed in the Austrian part of the eastern Alps (46°40′–47°35′ N, 9°50′–13°40′ E) where, in 1998, 910 glaciers covered a total area of 470 km^2 (Lambrecht and Kuhn, 2007). Figure 1 shows a map with the 1998 glacier coverage and an Advanced Spaceborne Thermal Emission and Reflection Radiometer (ASTER) digital elevation model (DEM) of elevations higher than 2000 m a.s.l. in the background. The ASTER DEM is for illustrative purposes; the analyses carried out in this study were performed with photogrammetrically derived DEMs. The largest part of the glacial area is along the main alpine divide, which follows the southern border of Austria from approximately 10°30′ E to 12°15′ E and continues eastward into the Austrian interior. Some small glaciers exist along the northern slopes of the Alps where peak altitudes are significantly lower. The ellipses indicate glacier groups of fairly uniform climatic and topographic settings (A–E in Fig. 1).

DATA AND METHODS

Two complete glacier inventories compiled in 1969 (Patzelt, 1980; Gross, 1987) and 1998 (Lambrecht and Kuhn, 2007; Kuhn and others, 2009a) are used in this study. Both surveys were conducted by aerial photogrammetry and include glacier boundaries as well as DEMs. Median elevations were calculated from the DEMs by linearly interpolating the area–elevation distribution, given in 50 m elevation bands, to the

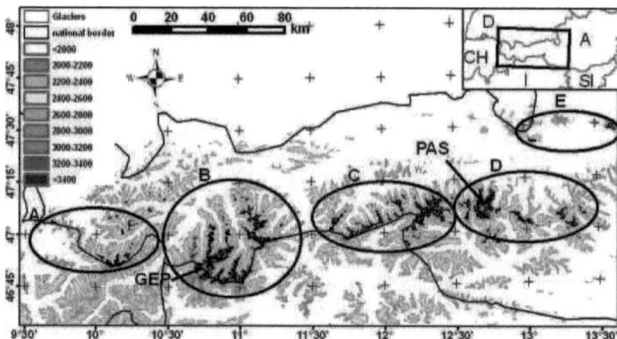

Fig. 1. Glacier cover in Austria according to the glacier inventory of 1998 (black outlines; Lambrecht and Kuhn, 2007) and an ASTER DEM in the background displaying elevations higher than 2000 m a.s.l. in colour. This figure distinguishes five areas (A–E) by their climatic and topographic settings. Austria's two largest glaciers, Pasterzenkees (PAS) and the Gepatschferner (GEP), are labelled.

elevation where 50% of the individual glacier's area is located above and 50% below. The median elevation is a characteristic glacier variable as shown by Kuhn and others (2009b) who use the median elevation to transfer mass-balance profiles from measured to unmeasured glaciers.

Mean glacier thickness changes, Δz, have been calculated as follows, where the overall volume change between 1969 and 1998 is $\Delta V_{1969-98}$ and the glacier areas in 1969 and 1998 are A_{1969} and A_{1998}:

$$\Delta \bar{z}_{1969-98} = \frac{\Delta V_{1969-98}}{\left(\frac{A_{1969}+A_{1998}}{2}\right)}. \tag{1}$$

Estimates of the accuracy of area and volume changes are made by Gross (1987), Eder and others (2000) and Lambrecht and Kuhn (2007).

We were interested in calculating a single temperature variable and single precipitation variable for each glacier that would best represent the local climate. As glacier mass balance in the Alps is influenced most strongly by summer temperature and winter precipitation (e.g. Kuhn and others, 1999), these two quantities were derived as a basis for comparisons.

The mean summer temperature was defined as the temperatures averaged between June and August inclusive. The temperature data were obtained from the European Centre for Medium-Range Weather Forecasts re-analysis project (ERA-40) (Uppala and others, 2005), which includes a dynamically consistent three-dimensional gridded dataset combining a numerical weather forecast model, meteorological observations and data from satellites from 1957 to 2002. The gridspace of the ERA-40 data is 1.25°. As temperature is highly elevation-dependent and thus a sensitive parameter when comparing glaciers of different sizes in different regions (with different altitudes as a consequence) a glacier-relevant constant level of 2500 m a.s.l. was defined. This is a typical elevation of high summer ablation areas in the Alps (e.g. WGMS, 2008) and always falls between the ERA-40 pressure levels of 700 and 850 hPa. Six-hour temperatures and the respective geopotential for these levels at the ERA-40 gridpoints were used. These values then were interpolated to derive time- and space-dependent lapse rates of temperature. Using the temperature and elevation of the 700 hPa pressure level together with the derived lapse rates, the mean summer temperature at 2500 m a.s.l. was determined.

Efthymiadis and others (2006) created a precipitation dataset with monthly resolution and 10'-sized cells for the entire Alps by homogenizing weather station data from more than 150 stations. Winter precipitation was determined for each glacier by interpolating linearly between the respective gridpoints and summing monthly precipitation values between October and May inclusive. The precipitation dataset is created using valley stations only (stations below 1900 m; Auer and others, 2005) because of the notorious uncertainties of high-altitude precipitation measurements (e.g. Yang and others, 1998; Sevruk, 2004). As precipitation has a higher local (x, y, z) variance than temperature, the precipitation field used in this study is less accurate than the temperature field.

Both winter precipitation and summer temperatures have been averaged to find the widespread 'climatological mean' of 1961–90 and are referred to as mean winter precipitation and mean summer temperature. The 'climatological mean' was chosen in order to make our results comparable with other studies; however, as we are concerned only with how these data change qualitatively over time (i.e. anomalies), the use of a different period would not alter the results of the study but would instead shift the curves along the ordinate.

To analyse temporal changes in winter precipitation and summer temperature the cumulative anomalies of these variables were derived from the climatological mean from 1969 to 1998, the years in which the two glacier inventories were compiled.

RESULTS

Glacier distribution and climatic setting

Figure 2a displays the relationship between mean winter precipitation over the period 1961–90 and median glacier elevation from the inventory of 1998. Mean winter precipitation varies considerably in the Austrian Alps: the wettest areas receive nearly three times as much precipitation as the

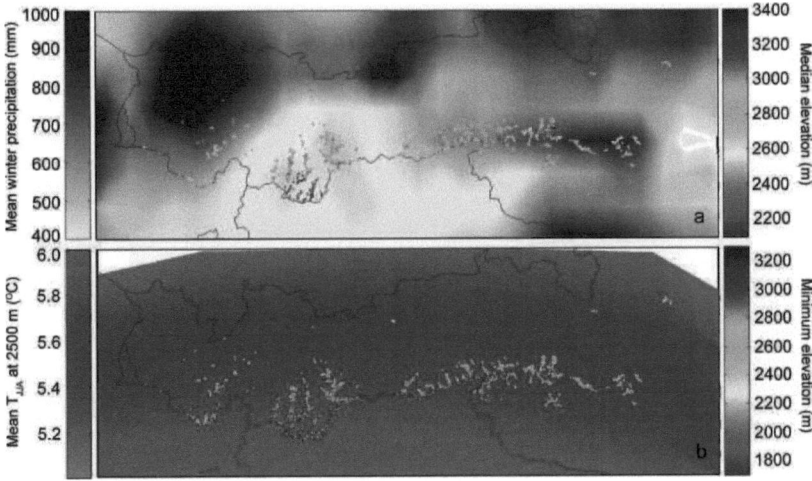

Fig. 2. (a) Mean winter precipitation (1961–90) in the study area (background) and each glacier's median elevation (circles). Precipitation data from Efthymiadis and others (2006). (b) Mean summer temperature (1961–90) at 2500 m a.s.l. derived from ERA-40 data (background) and each glacier's minimum elevation (circles).

driest. The drier the local climate, the earlier a glacier exposes bare ice at a given altitude and the longer the ablation period lasts. As a result, for a given altitude and during a period of general warming, glaciers situated in a drier climate are expected to lose more mass than glaciers situated in a wetter climate.

Mean winter precipitation is greatest in the easternmost and westernmost parts of Austria's glacial region and much lower in the centre. Glaciers in the north and northeast tend to be fairly wet as well. This pattern is readily explained by the fact that most winter precipitation falls out of weather systems from the north and northwest (Fliri, 1975). The inner alpine areas are shielded from the bulk of these systems by the northern ridges. The eastern ridges are less thoroughly shielded as the northern alpine ridge is more open there.

In general, the pattern of median elevations closely follows the pattern of winter precipitation. Wetter glaciers in the west and east have lower median elevations (e.g. 2500 m a.s.l. in the northwest), while glaciers in the dry Ötztal and Stubai region are typically much higher (~3000 m a.s.l.) (Kuhn and others, 2009b).

Temperature is highly elevation-dependent and thus the median elevation contains a simplified temperature signal. In turn, the median elevation of a glacier can be expected to be influenced by temperature conditions. To study the horizontal temperature signal, the mean summer temperature at 2500 m a.s.l. as derived from the ERA-40 is presented in Figure 2b. Summer temperatures at 2500 m a.s.l. average 5.5°C, with variations of <0.5°C in either direction. There is a north-to-south temperature gradient, with higher temperatures in the south.

For various size classes, the lowest altitude of each glacier in 1998 is related to mean winter precipitation and mean summer temperature at 2500 m a.s.l. in Figure 3. Each circle represents one glacier; the size of each circle is proportional to the area of the respective glacier. In each figure the correlation coefficient, r, between minimum elevations and mean winter precipitation is displayed. Correlations in Figure 3a–e are statistically significant at the 99% confidence level; the correlation in Figure 3f is statistically significant at the 95% confidence level. The relation between the parameters mentioned is investigated for all glaciers in Figure 3a and for individual size classes in Figure 3b–f.

Relationships exist between:

minimum elevation and mean winter precipitation. Figure 3a–f show generally higher values of minimum elevation in regions with less winter precipitation, resulting in negative slopes of the regressions. This supports our hypothesis that greater accumulation leads to more mass turnover and lower minimum elevations.

minimum elevation and mean summer temperature at 2500 m a.s.l. (Fig. 3, colour code). Areas with minimal elevations coincide with slightly higher temperatures than areas with lower minimum elevations. These also generally coincide with drier regions (compare with Fig. 2a and b).

minimum elevation and glacier area. In general, the larger the glacier, the lower the minimum elevation; this is due to larger mass turnover for large glaciers than for small glaciers. Note the progression of the cluster of dots from high (Fig. 3b) to low (Fig. 3f). For comparisons of glacier changes across different regions it is therefore advisable to compare glaciers of similar area only.

To quantify the relations described above objectively, a multiple regression was run, where the minimum elevation is z_{min} (m), the mean winter precipitation for 1969–90 is p_w (m), the mean summer temperature at 2500 m a.s.l. is T_{s2500}

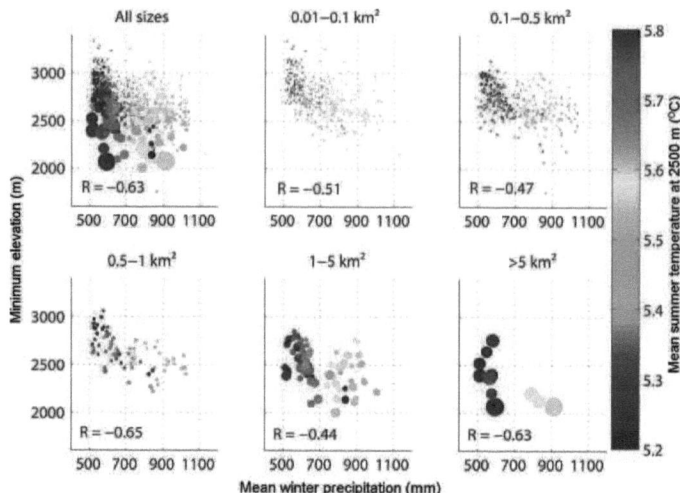

Fig. 3. Mean winter precipitation vs a glacier's minimum elevation, each point representing one glacier's climatic conditions. The size of the circles is proportional to the glacier's area and the colour code is a measure for the mean summer temperatures at 2500 m a.s.l. at the glacier's position. (a) All glaciers; (b–f) glaciers are divided into area classes.

(K) and the glacier area in 1998 is A_{1998} (m²):

$$z_{min} = -(9.86 \times 10^5) - (2.04 \times 10^2) \ln p_w + (1.76 \times 10^5) \ln T_{s2500} - 0.18 \times \sqrt{A_{1998}}. \quad (2)$$

The correlation coefficient between predicted and measured minimum elevations is 0.65. Logarithmic or square-root transformations were selected by evaluating different possibilities and choosing the best result (highest correlation coefficient). A similar regression analysis was performed with median elevations, but that leads to a weaker correlation ($r = 0.55$).

Glacier changes

In order to interpret the relationship between regional climate and resulting glacier changes, the spatial distributions of relative area changes and mean thickness changes between the glacier inventories of 1969 and 1998 are displayed in Figure 4. The overall results for the defined regions A–E are summarized in Table 1, wherein relative area changes and mean thickness changes of the regions are subdivided into glaciers smaller than 0.5 km² and glaciers larger than 0.5 km²; the total values are shown in the last two columns. Small glaciers generally shrank more in area and less in mean thickness than large glaciers. Regions A and E, the two regions situated furthest north in the study area, show less negative area changes than the more central alpine regions B, C and D, the only exception being relative area changes of glaciers larger than 0.5 km² in region A. With the same exception, changes within regions B and C are significantly more negative than within A and E. Reductions in area and mean thickness are of similar magnitude throughout regions B and C. Region D contains Austria's largest glacier, Pasterzenkees, which influences the negative record value for mean thickness changes of glaciers larger than 0.5 km² and is also visible in the negative record of mean thickness change values for all glaciers.

Small glaciers are strongly influenced by very local conditions such as wind, avalanches or topographic shading (e.g. Kuhn, 1995; DeBeer and Sharp, 2009). This makes it very difficult to interpret changes in the size or thickness of small glaciers as a direct result of climate change. In order to keep Figure 4a and b more comprehensible, changes of glaciers larger than 0.5 km² are displayed in colour; smaller glaciers are plotted as black crosses. Both relative area changes and mean thickness changes show a very heterogeneous pattern between the regions, as pointed out above and quantified in Table 1, but also contain gradients within the regions.

The westernmost glacier in region A shows a more negative value than the others in region A; however, the change in one rather small glacier should not be overinterpreted. More significantly, region B has a wide variety of values with a north-to-south gradient, which is visible both in relative area changes (Fig. 4a) and mean thickness change (Fig. 4b). At the southern edge of B, along the main alpine divide, glaciers shrank more and lost more of the initial thickness than in the north. Within region C, area changes tend to be less negative in the eastern part than in the western part; this is only barely visible in mean thickness changes. With the exception mentioned above, region D generally displays slightly less negative values than those in region E, where the western part (Hochköniggruppe) experienced significant area loss (approximately –20%). The eastern part (Dachstein) of region D changed little.

The pattern of mean thickness changes (Fig. 4b) is weakly but significantly related to the pattern of relative area changes, with a correlation coefficient of 0.29 at the 99% confidence level.

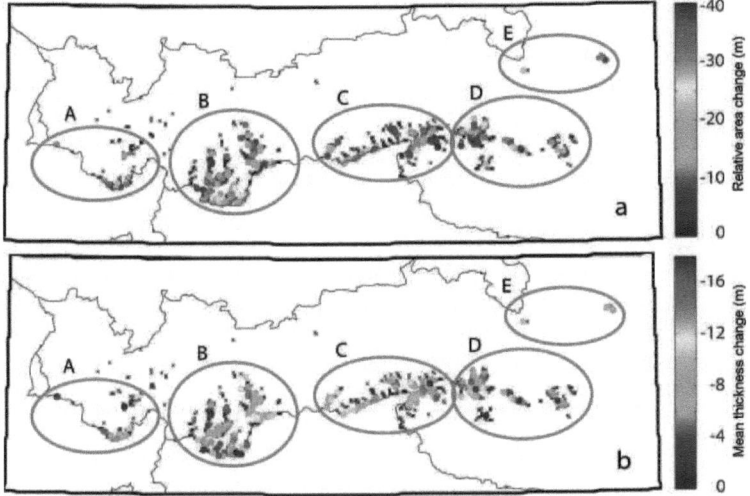

Fig. 4. (a) Relative area changes and (b) mean thickness changes between 1969 and 1998. The crosses display glacier coordinates of glaciers that are smaller than 0.5 km². The ellipses A–E show areas referred to in the text.

Climate change

In Figure 5, glacier area and volume changes are related to possible regional differences in climate change. Figure 5a displays the cumulative anomalies of winter precipitation from the long-term winter precipitation mean (1961–90) for the center of each region (Fig. 1). There is no significant overall trend of winter precipitation in regions A, B and E. Following a slightly negative precipitation anomaly in the early 1970s, eastern areas (D, E) show basically no trend from that date forward with the exception of a few years with positive anomalies in the early 1980s. In regions C and D, however, the study period also started with some negative anomalies (until the mid-1970s), followed by basically no anomalies until about 1985. After that, a series of years with stronger negative anomalies occurred again.

Temperature anomalies (Fig. 5b) show less spatial variability than do precipitation anomalies. There are six distinct time periods: an initial positive anomaly (1969–73); generally negative anomalies (1974–80); strong positive anomalies (1981–83); temperatures near the mean (1984–89); strong positive anomalies (1990–95); until 1998 nearly average ones. Both positive and negative temperature anomalies were strongest in region E and weakest in region A. There are no significant negative summer temperature anomalies after 1981.

DISCUSSION

The results presented above attempt to relate climatic parameters to typical descriptive glacier variables. We propose to invert this argument by asking: which climatic parameters are necessary to establish an environment in which a typical glacier can exist? While there are exceptions like Boggeneikees, which survives 5°C summer temperatures at <1750 m a.s.l. due to debris cover and a reliable ice supply through a steep narrow valley, glacier size is the

Table 1. Summary of relative area and mean thickness change for glaciers smaller than 0.5 km², glaciers larger than 0.5 km² and all glaciers in the areas shown in Figure 1 and the values for all glaciers in total

Region	<0.5 km²		>0.5 km²		All	
	ΔA	Δz	ΔA	Δz	ΔA	Δz
	%	m	%	m	%	m
A	−25.0	−8.2	−12.8	−9.7	−17.1	−9.2
B	−35.4	−8.7	−11.7	−10.1	−16.9	−9.8
C	−36.6	−8.0	−11.1	−10.1	−16.6	−9.7
D	−26.5	−7.5	−12.2	−11.4	−16.1	−10.4
E	−22.2	−6.7	−9.5	−7.0	−12.0	−6.9
Total	−32.7	−9.8	−11.6	−10.3	−16.6	−8.2

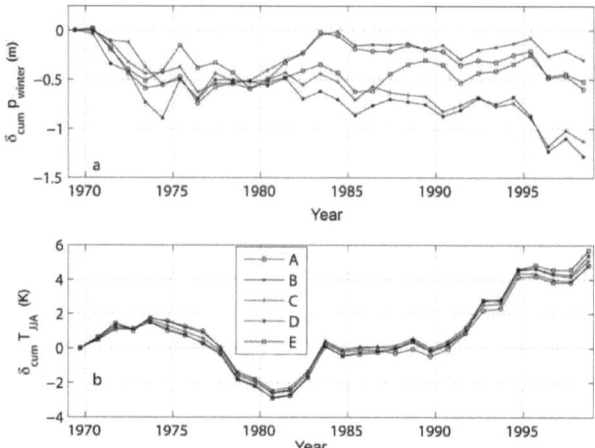

Fig. 5. Cumulative anomalies from the climatological mean (1961–90) at the centre of regions A–E (see Fig. 1). (a) Cumulative anomalies of winter precipitation and (b) cumulative summer temperature anomalies at 2500 m a.s.l.

dominant parameter that determines the minimum elevation (Fig. 3), suggesting that ice dynamics plays the most important role. The glacier size signal thus has to be removed in order to determine the effects of different climatic parameters on the minimum elevation; this is accomplished by comparing glaciers of similar size. The two largest glaciers, Pasterzenkees and the Gepatschferner (for locations see Fig. 1), have their minimum elevations at about the same altitude, ∼2080 m, despite very different precipitation conditions because of compensating differences in their respective topographic settings. The Gepatschferner has a larger accumulation-area ratio than Pasterzenkees because of larger portions of the glacier at high elevations. Therefore less accumulation is necessary in order to transport mass to similarly low elevations. The lowest elevation of these glaciers (2080 m) can be taken as the limit at which debris-free glaciers can exist in the study area. Smaller glaciers show a significant dependence of minimum elevations on mean winter precipitation (Fig. 3b–f).

While the sign of the correlation coefficient (negative in all correlations; Fig. 3a–f) demonstrates a dependence of minimum elevation on mean winter precipitation, the upper boundary of the point cloud, that is the highest elevation of the terminus for a given precipitation value, must be interpreted differently because the peak elevations of the glacier-covered areas vary widely within the study area.

The correlations shown in Figure 3a–f are rather weak. The wide scatter is due to the wide diversity of local topographic conditions and is unsurprising as it has been shown in various other studies how glaciers with similar climates can have very different dimensions and characteristics (e.g. Kuhn and others, 1985; Andreassen and others, 2008; Abermann and others, 2009).

The negative glacier changes between the two glacier inventories can be attributed to almost continuous positive temperature anomalies since 1981 throughout the study area, which is consistent with climate change interpretations made in globally directed studies (e.g. Solomon and others, 2007). Successive positive temperature anomalies amplify their own effect on a glacier's mass balance due to feedback mechanisms (e.g. albedo, heat conduction from surrounding terrain) as described in recent publications (e.g. Oerlemans and others, 2009; Fischer, 2010).

Individual years of positive mass and area changes in the late 1970s and early 1980s have been reported previously (Patzelt, 1985; Abermann and others, 2009). Figure 5 allows the attribution of these positive mass and area changes to changes in atmospheric conditions qualitatively: there was a sequence of negative temperature anomalies throughout the study area between 1974 and 1981 combined with some positive precipitation anomalies (Fig. 5).

Anomalies in winter precipitation have a less distinct temporal evolution than temperature anomalies. All regions show negative cumulative anomalies in total. While the period 1969–98 was marked by negative area and mass changes in Austrian glaciers, a sequence of positive anomalies of winter precipitation was observed (e.g. 1974 –83), especially in the western part of the country (regions A and B). The significantly more negative precipitation anomalies in regions C and D after 1983 add an argument for less negative glacier changes in A and B.

A reason why glacier changes in region B are more negative than, for example, in regions A and E, despite very similar precipitation and temperature anomalies, could be found by investigating the typical morphology of the glaciers in this region. Glaciers in region B consist of a relatively high fraction of low-lying (and comparably thick) valley glaciers that tend to react more strongly and on a longer timescale to climatically induced change. The extremely low glacier flow velocity values measured in recent years, for example on the Kesselwandferner (Abermann and others, 2007) and the Hintereisferner (Span and others, 1997), suggest an additional dynamic reason for the stronger recession: ice supply is reduced and the dynamically supported downwasting of these low-lying thick tongues follows as a direct consequence. Therefore, regionally varying temperature and precipitation anomalies alone cannot explain the regionally varying glacier change as shown in Figures 4 and 5.

Absolute values of mean precipitation should be interpreted with care because of the lack of altitudinal information in the dataset. However, in this study we base the line of argument on spatial differences of precipitation or temporal deviations from the mean and do not use absolute values for interpretation. It has been shown that anomalies of high and low elevations evolve similarly (Auer and others, 2005), which justifies the qualitative use of this dataset.

CONCLUSIONS

We have investigated the relationship between climatic parameters and glacier distribution in the Austrian Alps. Spatial differences in 1969–98 glacier changes can be attributed mainly to regional differences in local topography.

We found a relationship between glacier median elevation, glacier minimum elevation and mean winter precipitation. Spatial temperature gradients alone cannot explain the regionally very different glacier extents, but support the general pattern caused by spatial precipitation differences and general topographic conditions.

Since only basic glaciological data are required for this study, similar analyses are planned in other climatic regions. The inclusion of German, Swiss, Italian and Slovenian glaciers would allow an analysis of Alps-wide glacier changes and spatial gradients of their climatic setting. Meteorological data are also restricted to basic parameters: mean summer temperature including spatially varying lapse rates and mean winter precipitation which makes the analysis easy to apply over large areas.

Responding to the large glacier changes of the past decade, Abermann and others (2009) initiated a third lidar-based glacier inventory, which will soon be available as a database with which to continue investigations of regional climate change and its impact on glacier changes in the present and recent past.

ACKNOWLEDGEMENTS

This study was funded by the Commission for Geophysical Research, Austrian Academy of Sciences. We thank E. Dreiseitl, E. Schlosser and S. Kinter for comments and proofreading. We acknowledge I. Evans, an anonymous reviewer and the Editor for constructive comments.

REFERENCES

Abermann, J., H. Schneider and A. Lambrecht. 2007. Analysis of surface elevation changes on Kesselwand glacier – comparison of different methods. *Z. Gletscherkd. Glazialgeol.*, **41**, 147–167.

Abermann, J., A. Lambrecht, A. Fischer and M. Kuhn. 2009. Quantifying changes and trends in the glacier area and volume in the Austrian Ötzal Alps (1969–1997–2006). *Cryosphere*, **3**(2), 205–215.

Andreassen, L.M., F. Paul, A. Kääb and J.E. Hausberg. 2008. Landsat-derived glacier inventory for Jotunheimen, Norway, and deduced glacier changes since the 1930s. *Cryosphere*, **2**(2), 131–145.

Auer, I. *and 23 others*. 2005. A new instrumental precipitation dataset for the greater alpine region for the period 1800–2002. *Int. J. Climatol.*, **25**(2), 139–166.

Bolch, T., B. Menounos and R. Wheate. 2010. Landsat-based inventory of glaciers in western Canada, 1985–2005. *Remote Sens. Environ.*, **114**(1), 127–137.

DeBeer, C.M. and M.J. Sharp. 2009. Topographic influences on recent changes of very small glaciers in the Monashee Mountains, British Columbia, Canada. *J. Glaciol.*, **55**(192), 691–700.

Eder, K., R. Würländer and H. Rentsch. 2000. Digital photogrammetry for the new glacier inventory of Austria. *Int. Arch. Photogramm. Remote Sens.*, **33**(B4), 254–261.

Efthymiadis, D. *and 7 others*. 2006. Construction of a 10-min-gridded precipitation data set for the Greater Alpine Region for 1800–2003. *J. Geophys. Res.*, **111**(D1), D01105. (10.1029/2005JD006120.)

Evans, I.S. 2006. Glacier distribution in the Alps: statistical modelling of altitude and aspect. *Geogr. Ann., Ser. A*, **88**(2), 115–133.

Evans, I.S. and N.J. Cox. 2005. Global variations of local asymmetry in glacier altitude: separation of north–south and east–west components. *J. Glaciol.*, **51**(174), 469–482.

Evans, I.S. and N.J. Cox. 2010. Climatogenic north–south asymmetry of local glaciers in Spitsbergen and other parts of the Arctic. *Ann. Glaciol.*, **51**(55), 16–22.

Fischer, A. 2010. Glaciers and climate change: interpretation of 50 years of direct mass balance of Hintereisferner. *Global Planet. Change*, **71**(1–2), 13–26.

Fliri, F. 1975. *Das Klima der Alpen im Raume von Tirol*. Innsbruck, Universitätsverlag Wagner. (Monographien zur Landeskunde Tirols 1.)

Gross, G. 1987. Der Flächenverlust der Gletscher in Österreich 1850–1920–1969. *Z. Gletscherkd. Glazialgeol.*, **23**(2), 131–141.

Haeberli, W. and M. Hoelzle. 1995. Application of inventory data for estimating characteristics of and regional climate-change effects on mountain glaciers: a pilot study with the European Alps. *Ann. Glaciol.*, **21**, 206–212.

Hoelzle, M., T. Chinn, D. Stumm, F. Paul and W. Haeberli. 2007. The application of glacier inventory data for estimating past climate change effects on mountain glaciers: a comparison between the European Alps and the Southern Alps of New Zealand. *Global Planet. Change*, **56**(1–2), 69–82.

Kääb, A., F. Paul, M. Maisch, M. Hoelzle and W. Haeberli. 2002. The new remote-sensing-derived Swiss glacier inventory: II. First results. *Ann. Glaciol.*, **34**, 362–366.

Knoll, C., H. Kerschner and J. Abermann. 2009. Development of area, altitude and volume of South Tyrolean glaciers since the Little Ice Age maximum. *Z. Gletscherkd. Glazialgeol.*, **42**(1), 19–36.

Kuhn, M. 1995. The mass balance of very small glaciers. *Z. Gletscherkd. Glazialgeol.*, **31**(1–2), 171–179.

Kuhn, M., G. Markl, G. Kaser, U. Nickus, F. Obleitner and H. Schneider. 1985. Fluctuations of climate and mass balance: different responses of two adjacent glaciers. *Z. Gletscherkd. Glazialgeol.*, **21**(1–2), 409–416.

Kuhn, M., E. Dreiseitl, S. Hofinger, G. Markl, N. Span and G. Kaser. 1999. Measurements and models of the mass balance of Hintereisferner. *Geogr. Ann.*, **81A**(4), 659–670.

Kuhn, M., A. Lambrecht, J. Abermann, G. Patzelt and G. Gross. 2009a. *Die österreichischen Gletscher 1998 und 1969, Flächen- und Volumenänderungen*. Wien, Österreichische Akademie der Wissenschaften. (Landesverteidigung Projektbericht 10.)

Kuhn, M., J. Abermann, M. Bacher and M. Olefs. 2009b. The transfer of mass-balance profiles to unmeasured glaciers. *Ann. Glaciol.*, **50**(50), 185–190.

Lambrecht, A. and M. Kuhn. 2007. Glacier changes in the Austrian Alps during the last three decades, derived from the new Austrian glacier inventory. *Ann. Glaciol.*, **46**, 177–184.

Oerlemans, J., R.H. Giesen and M.R. van den Broeke. 2009. Retreating alpine glaciers: increased melt rates due to accumulation of dust (Vadret da Morteratsch, Switzerland). *J. Glaciol.*, **55**(192), 729–736.

Patzelt, G. 1980. The Austrian glacier inventory: status and first results. *IAHS Publ.* 126 (Riederalp Workshop 1978 – *World Glacier Inventory*), 181–183.

Patzelt, G. 1985. The period of glacier advances in the Alps: 1965 to 1980. *Z. Gletscherkd. Glazialgeol.*, **21**(1–2), 403–407.

Schiefer, E. and B. Menounos. 2010. Climatic and morphometric controls on the altitudinal range of glaciers, British Columbia, Canada. *Holocene*, **20**(4), 517–523.

Sevruk, B. 2004. *Niederschlag als Wasserkreislaufelement: Theorie und Praxis der Niederschlagsmessung.* Zürich, Nitra.

Shea, J.M., S.J. Marshall and J.M. Livingston. 2004. Glacier distributions and climate in the Canadian Rockies. *Arct. Antarct. Alp. Res.*, **36**(2), 272–279.

Solomon, S. *and 7 others, eds.* 2007. *Climate change 2007: the physical science basis. Contribution of Working Group I to the Fourth Assessment Report of the Intergovernmental Panel on Climate Change.* Cambridge, etc., Cambridge University Press.

Span, N., M. Kuhn and H. Schneider. 1997. 100 years of ice dynamics of Hintereisferner, central Alps, Austria, 1894–1994. *Ann. Glaciol.*, **24**, 297–302.

Uppala, S.M. *and 45 others.* 2005. The ERA-40 re-analysis. *Q. J. R. Meteorol. Soc.*, **131**(612), 2961–3212.

World Glacier Monitoring Service (WGMS). 2008. *Fluctuations of glaciers 2000–2005 (Vol. IX)*, *ed.* Haeberli, W., M. Zemp, A. Kääb, F. Paul and M. Hoelzle. ICSU(FAGS)/IUGG(IACS)/UNEP/UNESCO/WMO, World Glacier Monitoring Service, Zürich.

Yang, D. *and 6 others.* 1998. Accuracy of NWS 8-inch standard nonrecording precipitation gauge: results and application of WMO intercomparison. *J. Atmos. Oceanic Technol.*, **15**(1), 54–68.

A2: PAPER II: ON THE POTENTIAL OF VERY HIGH-RESOLUTION DEMS IN GLACIAL AND PERIGLACIAL ENVIRONMENTS

By: Abermann, J., A. Fischer, A. Lambrecht and T. Geist. *Published in The Cryosphere*, **4**(1): 53-65.

The Cryosphere

On the potential of very high-resolution repeat DEMs in glacial and periglacial environments

J. Abermann[1], A. Fischer[2], A. Lambrecht[2], and T. Geist[3]

[1]Austrian Academy of Sciences, Commission for Geophysical Research, Vienna, Austria
[2]Institute of Meteorology and Geophysics, University of Innsbruck, Innsbruck, Austria
[3]FFG – Austrian Research Promotion Agency/ALR – Aeronautics and Space Agency, Vienna, Austria

Received: 9 June 2009 – Published in The Cryosphere Discuss.: 1 July 2009
Revised: 29 December 2009 – Accepted: 12 January 2010 – Published: 25 January 2010

Abstract. The potential of high-resolution repeat DEMs was investigated for glaciological applications including periglacial features (e.g. rock glaciers). It was shown that glacier boundaries can be delineated using airborne LIDAR-DEMs as a primary data source and that information on debris cover extent could be extracted using multi-temporal DEMs. Problems and limitations are discussed, and accuracies quantified. Absolute deviations of airborne laser scanning (ALS) derived glacier boundaries from ground-truthed ones were below 4 m for 80% of the ground-truthed values. Overall, we estimated an accuracy of +/−1.5% of the glacier area for glaciers larger than 1 km². The errors in the case of smaller glaciers did not exceed +/−5% of the glacier area. The use of repeat DEMs in order to obtain information on the extent, characteristics and activity of rock glaciers was investigated and discussed based on examples.

1 Introduction

Glacial and periglacial environments have been changing rapidly in the past decades (e.g., Dyurgerov and Meier, 2000; Haeberli, 1999) as a result of climate change (e.g., Lemke et al., 2007; Trenberth et al., 2007). Determination of their actual geometry is crucial to the monitoring process and for understanding the short-term responses of their extent.

The mapping of glacier extent and volume changes using remote sensing techniques is a widely used and powerful tool. Various studies show both the potential and the limitations of using satellite data (e.g., Andreassen et al., 2008; DeBeer and Sharp, 2007; Paul et al., 2007; Kääb et al., 2002), airborne techniques such as photogrammetry (e.g., Würländer et al., 2004; Patzelt, 1980) or LIDAR (light detection and ranging, e.g., Geist and Stötter, 2007; Geist et al., 2003; Favey et al., 2002; Baltsavias et al., 2001a). Automatic or semi-automatic classification algorithms (Hendriks and Pellikka, 2007; Höfle et al., 2007; Kodde et al., 2007; Paul et al., 2002; Rott and Markl, 1999) are used to classify glacier areas.

However, the mapping of debris-covered glacier areas proves to be problematic in the case of both automatic and manual methods (e.g., Knoll and Kerschner, 2009; Paul et al., 2002; Hendriks and Pellikka, 2007). Fig. 1 illustrates the problem of glacier boundary delineation of complex glacier boundaries, addressed further on. Neither the highly resolved orthophotograph (Fig. 1b: 0.5 m spatial resolution) nor the RGB-composite of a Landsat scene of the Hintereisferner glacier margin (Fig. 1b: bands 4, 5 and 6 as proposed by Rott and Markl, 1999) allow for definite detection of the glacier boundary, including the debris-covered part of the glacier tongue as demonstrated later. Furthermore, the automatic mapping of small glaciers is difficult in particular from space (e.g., Paul et al., 2002). Lambrecht and Kuhn (2007) showed that 79% of all Austrian glaciers are smaller than 0.5 km², and 43% are smaller than 0.1 km².

The main aim of this study is thus to investigate the potential of a set of high-resolution DEMs to map glacier extents with sufficient accuracy, independent of their size and debris cover. Accuracies are quantified and limitations discussed. Further, we highlight the potential of repeat DEMs for interpretation of small-scale rock glacier surface elevation changes.

Correspondence to: J. Abermann
(jakob.abermann@uibk.ac.at)

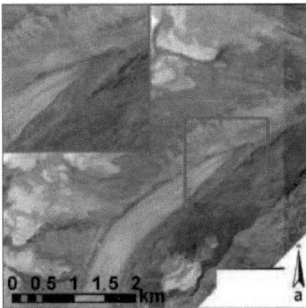
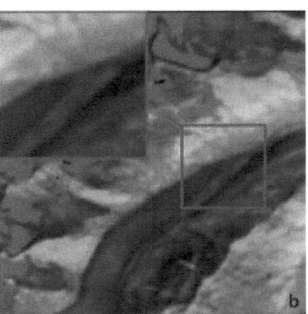

Fig. 1. (a) The orthophotograph (0.5 m spatial resolution) taken 2003, and (b) the Landsat 7 ETM+ scene (taken 10 September 2004) of the area around Hintereisferner, with the channels 4, 5, 6 H as an RGB-composite. The red rectangle indicates the extent of the close-up in the upper left corner.

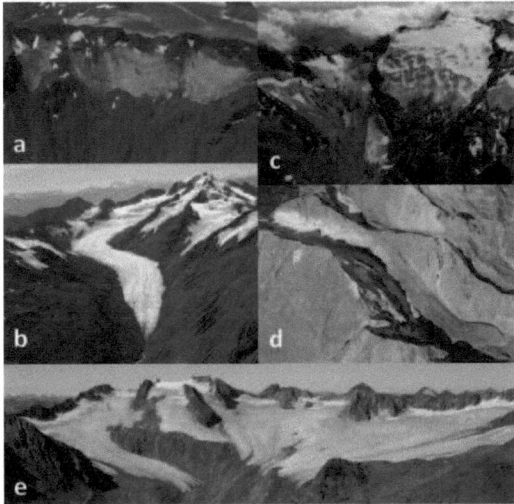

Fig. 2. Aerial photographs of (a) Mittlerer Guslarferner (9 September 2008), (b) Hintereisferner (9 September 2008), (c) Rotmoos- and Wasserfallferner (28 August 2009) and (e) Vernagtferner (28 August 2009). (d) is an oblique perspective of a SPOT-image of Reichenkar rock glacier presented in Google Earth (Google Earth, 2009).

2 Test sites and data

Four glaciers and one rock glacier in the Ötztal Alps and the neighboring part of the Stubai Alps (ca. 47° N, 11° E) were chosen as test sites. Aerial photographs in Fig. 2 provide a view of the individual study sites. Local climatic conditions during the past four decades are described in Abermann et al. (2009). To test the glacier boundary delineation for a small, debris-free glacier, we chose Mittlerer Guslarferner (Fig. 2a). Nearby Hintereisferner (Fig. 2b) has been the object of extensive glaciological investigations over the past decades. This has resulted in a large number of DEMs,

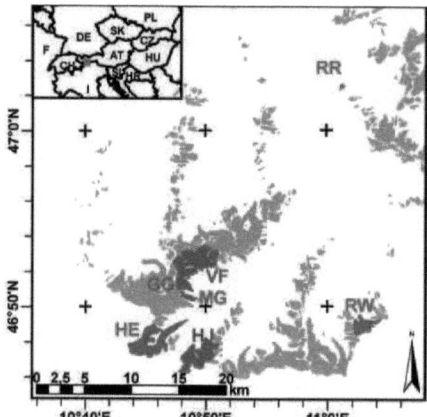

Fig. 3. The study area: Glaciers in the Ötztal and Stubai Alps (grey) with three reference glaciers (HE: Hintereisferner, MG: Mittlerer Guslarferner, RW: Rotmoos- und Wasserfallferner) and one rock glacier (red, RR: Reichenkar rock glacier). Ground truth has been performed for the blue glaciers (GG: Großer Guslarferner, HJ: Hochjochferner and VF: Vernagtferner) as well as Hintereisferner.

Table 1. Summary of technical specifications of the LIDAR acquisition campaign of the regional government of Tyrol, 2006.

Sensor	Optech ALTM3100
Laser Wavelength	1064 nm
Scan Frequency	33 Hz
Scan angle	+/−20°
Point density	Minimum: 1 point/4 m^2
Measurement frequency	71 kHz
Average flight height	1100 m a.g.l.
Mode	Last Pulse
Interpolation software	SCOP++
Interpolation method	Moving Planes
Horizontal accuracy	+/−0.3 m
Vertical accuracy	+/−0.1 m
Spatial resolution	1 m

remote sensing data and ground truth. For this reason, Hintereisferner was chosen for the investigation of the performance of the data set on a debris-covered margin. Problems with glacier boundary delineations in firn areas are highlighted, using Rotmoosferner as an example (Fig. 2c); and ground truthing is demonstrated at Vernagtferner (Fig. 2e) and was performed on three more glaciers. The thoroughly investigated ice-cored Reichenkar rock glacier (Krainer et al., 2002; Krainer and Mostler, 2000, Fig. 2d) is used as an example to show potential applications for delineating rock glacier extents. Figure 3 shows a map of the whole study area with glaciers in the Ötztal and Stubai Alps (grey) and the example glaciers (red).

For the purpose of ground control, geodetically measured points are included. These were acquired with a theodolite and an electro-optical rangefinder (Kern, DM501) achieving an xyz-accuracy of less than 5 cm standard error (H. Schneider, personal communication, 2009). The results of the length change measurements are published annually by the Austrian Alpine Club (issues up to and including 2003/2004 named "Mitteilungen des Österreichischen Alpenvereins", after 2004/2005 named "Bergauf"; Patzelt, 2005; Patzelt, 2006). Within the study area these comparably accurate length change measurements were performed at the margins of Hintereisferner, Vernagtferner, Großer Guslarferner and Hochjochferner, totalling 118 measured points compared using an independent method. These glaciers are colored blue in Fig. 3. We show the example of Vernagtferner in Sect. 4.5. Coincidentally these ground-based measurements, which were taken on 22 and 23 August 2006, deviate from the LIDAR acquisition date (23 August 2006) by one day at most.

DEMs with 10 m spatial resolution acquired in 1997, and high-resolution LIDAR-DEMs from 2006, were available for all test sites. The DEMs from 1997 were acquired during the compilation of the second Austrian glacier inventory using digital photogrammetry (Kuhn et al., 2009; Lambrecht and Kuhn, 2007; Würländer and Eder, 1998). The 2006 LIDAR-DEMs were acquired by the regional government of Tyrol. The technical specifications of this LIDAR acquisition campaign are summarized in Table 1.

Another set of LIDAR-DEMs covers a study area around Hintereisferner for which 14 DEMs were produced between 2001 and 2007. Relative horizontal accuracies are better than 1 m and relative vertical accuracies better than 0.3 m according to Geist and Stötter (2007), where more technical specifications of this acquisition campaign are described. For the application of our method three survey flights were chosen since they were acquired at a similar time of year (October 2001, 2004 and 2005) close to the minimum snow extent.

The example of Hintereisferner has been used to demonstrate the applicability and potential of the repeat high-resolution DEMs as shown in Fig. 1. The orthophotograph (Fig. 1a) was acquired in the OMEGA project (e.g. Pellikka and Rees, 2009; Kuhn, 2007). The Landsat scene was acquired on 10 September 2004 (path 193, row 27, id LE71930272004254ASN01). Table 2 shows details on the acquisition dates of all data used, its accuracies as well as image and spatial resolution.

Table 2. Summary of the acquisition dates as well as resolutions and accuracies of the data used in this study.

Source data	Date	Band	Image resolution [m]	DEM resolution [m]	Horizontal accuracy [m]	Vertical accuracy [m]
Aerial photography	11 September 1997		1	10	1	0.7
LIDAR	11 October 2001		–	1	< 1	< 0.3
LIDAR	5 October 2004		–	1	< 1	< 0.3
LIDAR	12 October 2005		–	1	< 1	< 0.3
LIDAR	23 August 2006		–	1	0.3	0.1
Landsat	10 September 2004	4,5,6 H	30	–	ca. 30	–
Aerial photography	12 August 2003		0.5	–	1	–

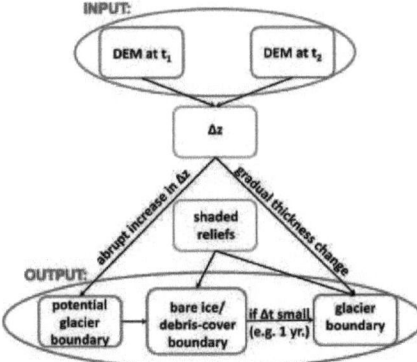

Fig. 4. Workflow of the glacier mapping procedure with multi-temporal (high-resolution) DEMs.

3 Methodology

Glacier boundary delineation was performed following the schematic workflow as sketched in Fig. 4, where t_1 is the point in time of the first and t_2 of the second DEM. To distinguish between glaciers and their surroundings, surface elevation changes between the two points in time were calculated. Investigating the course of surface elevation changes provided information on the actual glacier boundary as well as on debris-cover or dead-ice extent: a gradual increase of surface elevation loss from the glacier margin at t_1 upwards indicated a debris-free glacier tongue with a maximal change at the position of the glacier margin of t_2.

A glacier with debris cover evolves differently from bare ice due to the fact that thick debris cover reduces ablation (Kirkbride and Warren, 1999). For this reason, elevation differences between t_1 and t_2 are significantly smaller at the debris-covered parts and show instant increase at the place where debris cover meets bare ice. We used these differences to gain information about the occurrence and, depending on the time interval between the acquired DEMs, the extent of debris cover. Examples of these cases will follow in Sect. 4.

Two relief-shaded representations of the high-resolution DEM (in the following: shaded reliefs) at t_2 with different azimuth angles for illumination (315° and 135°) were calculated to present optimal visualization of contrasts in different aspects. Taking advantage of the already existing glacier inventory from a previous date (Lambrecht and Kuhn, 2007), we then undertook a qualitative analysis of the ways in which ice thickness has evolved from the former glacier terminus position upwards. Indicating the existence of former glacier boundaries is not mandatory but it saves time, as these show where to expect glacier-covered areas. Nevertheless, even if a former data set of glacier boundaries is found to exist, it is advisable to test this using a difference raster, in order to prevent a glacier not captured in a previous study from escaping capture in a new one.

If a gradual increase in ice thickness loss from the former glacier margin upwards was detected, we set the glacier boundary directly by manually digitizing the strongest roughness change in the shaded reliefs via visual inspection. Upon detection of an abrupt increase in ice thickness loss, shaded reliefs were used to establish the boundary between bare ice and dead ice or debris-covered ice. The difference raster helped to determine the extent of the debris-covered areas in cases where the temporal resolution was high enough (e.g. 1 year). In cases where temporal resolution was lower (e.g., years to decades), the potential glacier boundary could be derived in areas where a significant thickness change had occurred.

Surface elevation changes are much smaller in accumulation zones of glaciers (Abermann et al., 2009). We therefore could only partly take advantage of the difference raster for boundary delineation and thus used the roughness changes in the shaded reliefs. If they were not distinct enough, we used orthophotographs to map the glacier extent in these areas.

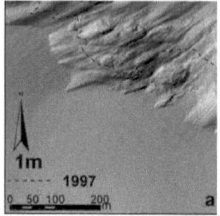 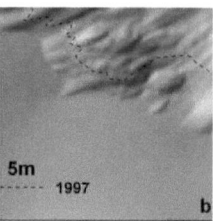 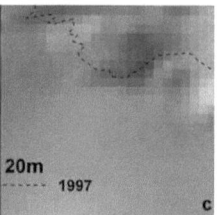 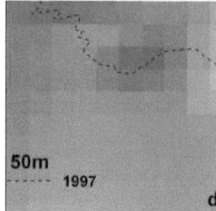

Fig. 5. Shaded reliefs of Mittlerer Guslarferner's margin calculated out of the LIDAR-DEMs with (**a**) the original 1 m and the resampled (**b**) 5 m, (**c**) 20 m and (**d**) 50 m resolutions. The glacier margin of 1997 is dashed in all figures; the glacier margin of 2006 is not displayed in order to show the distinct roughness differences between rock and ice objectively.

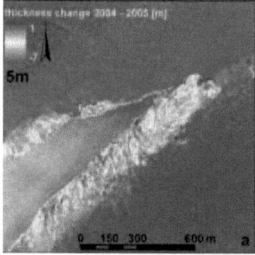 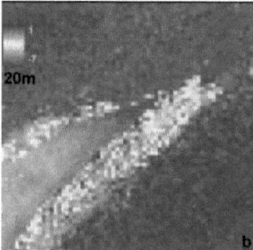 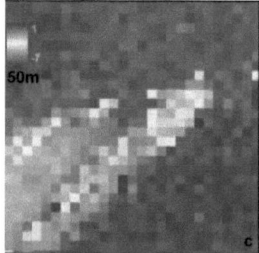

Fig. 6. Ice thickness changes between 2004 and 2005 using (**a**) 5 m spatial resolution, (**b**) 20 m resolution and (**c**) 50 m resolution for the tongue of Hintereisferner.

The influence of the spatial resolution of the DEM on the quality of the glacier boundary delineation with high-resolution DEMs as a main data source is highlighted in Figs. 5 and 6. We calculated four shaded reliefs out of differently resampled DEMs (Fig. 5a–d) for a part of the margin of Mittlerer Guslarferner. A qualitative analysis of surface roughness differences is applicable for DEMs that exist at a resolution better than 5 m. One-meter DEMs are optimal and allow the use of orthophotographs or any other additional information for glaciers without debris cover to be omitted (Fig. 5a). A spatial resolution of 20 m or higher fails to resolve roughness changes adequately (Fig. 5c and d).

Figure 6 shows as an analogy the ice thickness changes calculated from differently resampled DEMs at the margin of Hintereisferner (same extent as in Fig. 1). The differences between the rocky surroundings, the debris-covered part of the margin and the debris-free ice is visible up to the 50 m resolution. However, since the differences between the surface characteristics are small (compare noise in rocky surroundings with debris-covered part in Fig. 6b and c), no significant conclusions can be drawn for spatial resolutions larger than 5 m. Thus the proposed method is limited to high-resolution remote sensing data.

While investigating the potential of a sequence of high-resolution DEMs for rock glacier monitoring we were limited to an example where a photogrammetric DEM of 1997 and the LIDAR-DEM of 2006 exist (Reichenkar rock glacier) due to the lack of repeat LIDAR-DEMs. Surface elevation changes were calculated and their distribution analyzed in order to detect the margin of the rock glacier as well as the results from small- and larger-scale dynamics.

4 Results

We now highlight the results of the glacier boundary delineation, providing reference glaciers of different characteristics as examples (see 4.1.–4.4). The course of elevation changes from the former glacier boundary upwards is presented visually by profiles that are indicated on the maps and displayed as inserted figures.

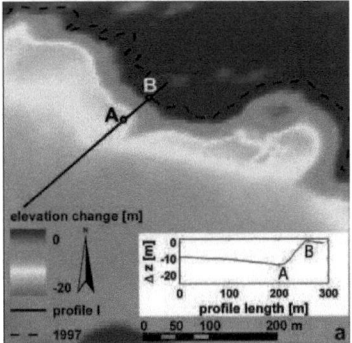 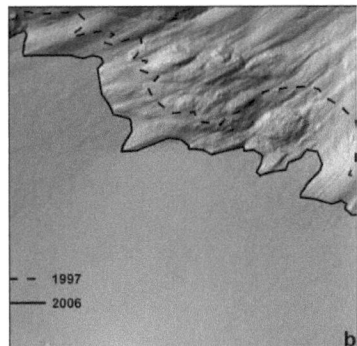

Fig. 7. (a) Elevation change 1997–2006 with the glacier boundary of 1997 (dashed) on Mittlerer Guslarferner and a profile of surface elevation changes inserted in the lower right corner. In (b) the shaded relief of the LIDAR-DEM 2006 is shown with the glacier boundary of 2006 (solid line). The same scene as in Fig. 5 is shown.

4.1 Debris-free glacier tongue: the example of Mittlerer Guslarferner

The small (0.5 km^2) debris-free Mittlerer Guslarferner had a gradual loss in ice thickness from the former glacier margin (B) upwards (Fig. 7a, same extent as Fig. 5). Between B (glacier margin in 1997) and A (glacier margin in 2006), absolute values of surface elevation changes increased gradually up to about 13 m at A and decreased from A upwards, as shown in the inserted profile. An optimal delineation of the glacier extent was performed by following the pronounced roughness changes in the shaded reliefs as shown in Fig. 7b. Compare also Fig. 5, where the delineated glacier boundary was not added and thus the roughness changes are more clearly visible.

4.2 Accumulation zone: the example of Rotmoosferner

We achieved good results in large parts of the accumulation area by qualitatively analyzing the roughness changes in the shaded reliefs. In this way the glacier surface could be distinguished from its rocky surroundings. As suggested in UNESCO (1970) as well as Paul et al. (2009), we included adjacent perennial snow-covered areas in the glacier surface area. The acquisition dates of the LIDAR-DEMs (October and late August, see Table 2) were optimal: they were close to the minimum snow extent in the Alps when the surroundings of the glaciers were likely to be snow-free. In some cases also in the lower parts of the accumulation zone, our analysis of surface elevation changes helped us to decide which areas to include in the glacier extent. We also performed a qualitative analysis of aerial photographs which helped us reach conclusions about the remaining somewhat ambiguous areas. Fig. 8a provides an example of the firn area on Rotmoosferner. It was not possible, by analyzing the shaded relief of the DEM (black ellipse) alone, to tell whether it was partly debris-covered ice or consisted only of rocks. Our analysis of the surface elevation changes failed to yield a distinct answer since surface elevation changes were very minimal in this region. Fortunately, a qualitative comparison with an aerial photograph taken by the regional government of Tyrol in 2003 (TirisMaps, 2009) provided a strong indication. Crevasse patterns could be seen in this debris- or rock-covered part of the glacier (Fig. 8b).

4.3 Debris-covered glacier margins: the example of Hintereisferner

In cases where we identified an abrupt increase in elevation loss around the former glacier boundary, we followed a different workflow from the one in Fig. 4. An example of this is given in Fig. 9 for the margin of Hintereisferner (same extent as in Fig. 6). Figure 9a shows the differences calculated between 2001 and 2005. As UNESCO (1970) as well as Paul et al. (2009) suggested, adjacent debris-covered areas and dead-ice bodies have to be included in glacier inventories. In accordance with this, the areas in which a significant elevation change had occurred were included in the so-called "potential" glacier area (Fig. 9b). The significance of the potential glacier area depends on the temporal resolution of the multi-temporal DEMs. If the two DEMs used were acquired at widely differing points in time (e.g. decades), and if during this period a significant ice volume loss occurred, it may well be that ice that was stored beneath the debris cover was partly melted by the time of the second acquisition date. In such cases the additional use of multi-temporal DEMs should provide a clue as to where ice could still exist

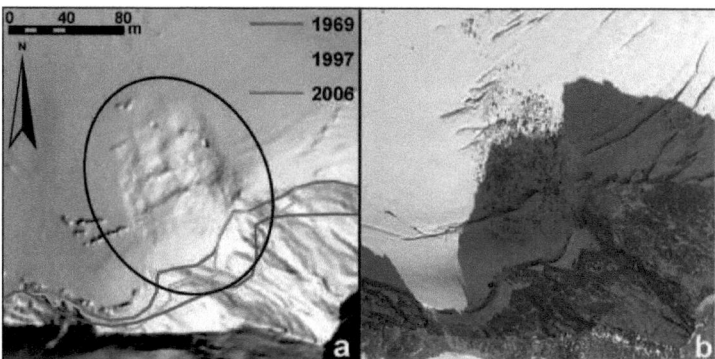

Fig. 8. (a) Rotmoosferner's firn area displayed as a shaded relief of the 2006 DEM with glacier margins of 1969 (blue), 1997 (orange) and 2006 (red) and the ambiguous area (black ellipse) as addressed in the text. **(b)** shows an aerial photograph of 2003.

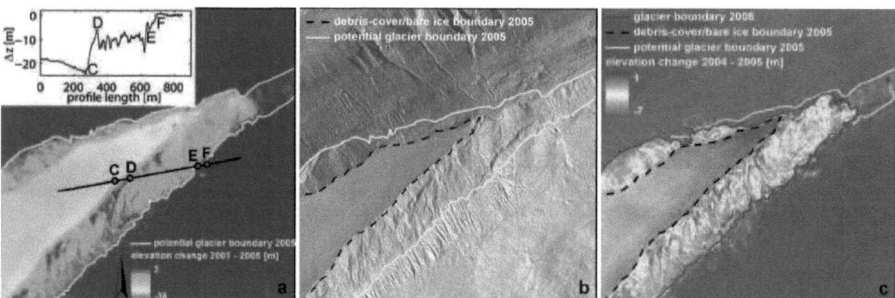

Fig. 9. In **(a)** the ice thickness changes between 2001 and 2005 are shown together with the potential glacier boundary (grey). A profile of surface elevation changes is inserted in the upper left corner. The calculated shaded relief from the 2005 DEM is shown in **(b)**, including the debris-cover/bare ice boundary. Indicator 1 marks gully-formation that is discussed in Sect. 5. In **(c)**, DEM-differences between 2004 and 2005 allow for the detection of the glacier boundary 2005.

below the debris cover. If the interval between the two DEMs is short (e.g. years, not decades), we conclude that in places where an elevation change has occurred, it may be assumed that there is still ice below. In our study area around Hintereisferner we had the advantage of a very good temporal resolution; therefore, the glacier extent could be determined very precisely using two DEMs with a one-year time difference (Fig. 9c).

The surface elevation changes along the profile in Fig. 9a allowed for a more detailed analysis comparing bare ice and debris-covered areas. From the profile starting point until C, absolute values of surface elevation changes increased gradually (as compared to the example in Fig. 7) to more than 20 m, since this is the bare-ice region. C coincides with the debris-cover/bare-ice boundary (Fig. 9b). Within a short distance only, between C and D, surface elevation changes decreased rapidly to around 10 m. This transition between debris-free and debris-covered ice can follow a less-pronounced shape, depending on the thickness of the debris layer and the topography of the surroundings. A decrease in thickness changes compared to the bare ice is to be expected. The insulation of the debris between D and E thus reduced ablation by about half for the example given. Thickness variations in the debris layer resulted in significant ablation variations on a smaller scale (e.g., meters to tens of meters). Between E and F the debris cover thickened and the transition from debris-covered ice to ice-free surroundings took place. F coincided with the potential glacier extent.

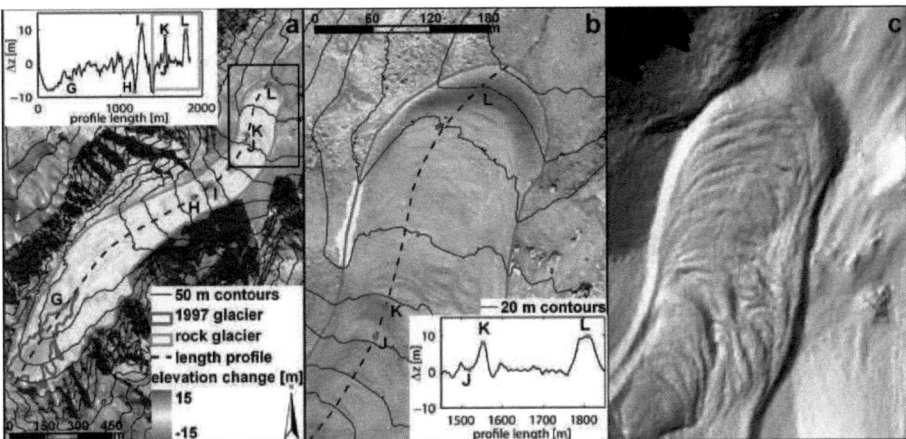

Fig. 10. Reichenkar rock glacier at 47°03′56″ N, 11°02′56″ E: **(a)** shows an overview of the area around Reichenkar rock glacier with the glacier boundary 1997 in blue and the rock glacier's extent of 2006 in orange. Surface elevation changes are shown in a color scheme and the elevation changes along the length profile (dashed in the map) are plotted in the inserted rectangle (upper left corner). The black rectangle on the upper right marks the extent of **(b)** and **(c)**. **(b)** shows surface elevation changes with the same color scheme as in (a) with 50% transparency and the orthophotograph of 1997 below. **(c)** displays a shaded relief calculated from the LIDAR-DEM 2006.

4.4 Potential for rock glacier monitoring: Reichenkar rock glacier

Another interesting mode of application of LIDAR and multi-temporal DEMs is the mapping and monitoring of rock glaciers (as mentioned in the literature, e.g. Kääb, 2008a). Figure 10a shows Reichenkar rock glacier with the orthophotograph of 1997 in the background and surface elevation changes (1997–2006) in a color code. Over large parts of the rock glacier a surface elevation loss occurred between 1997 and 2006 due to the comparably thick ice core of the rock glacier (between 30 and 40 m (Krainer et al., 2002) that was gradually melting out. The most negative values were in the uppermost part (upwards from G) where a rock-free glacier (Reichenkarferner, blue line, Lambrecht and Kuhn, 2007) transforms into the rock glacier. The longitudinal profile inserted in Fig. 10a reveals an interesting feature at markers H and I, where an elevation loss at H is followed by an elevation gain further downstream (I). This could be attributed to the propagation of a ridge downwards, a small-scale phenomenon connected with rock glacier elevation changes that is also found in other studies (e.g. Kääb and Vollmer, 2000). Even smaller-scale change patterns were found further down the longitudinal profile, where a significant surface elevation gain is visible at K (up to 8 m; the close-up of this region is found in Fig. 10b with a zoom into the profile inserted). Hardly any change in elevation was observed upwards from K (e.g. at J) indicating a compressive flow of the rock glacier due to a change in surface slope from steeper (upwards from K) to flatter (from K downwards) terrain. In this part of the rock glacier, considerable velocities of up to $4\,\mathrm{m\,a^{-1}}$ were measured by Krainer and Mostler (2006). A pronounced positive surface elevation change occurred at the rock glacier's snout with up to 10 m surface elevation gain at L This accompanied an advance of the snout of about 25 m in 9 years. This result fits well to measured horizontal flow velocities between 2.5 and $2.9\,\mathrm{m\,a^{-1}}$ at the central part of the rock glacier's snout as measured by Krainer and Mostler (2006). The distinct snout of the rock glacier can be delineated very well manually following the pronounced roughness changes of the shaded relief (Fig. 10c).

4.5 Accuracy and ground truth

The accuracy estimation of the proposed method was achieved in different ways. Given that the method is based on a manual delineation, we had to anticipate and deal with interpretation errors. To this end, we first compared the results obtained by two different parties for certain glaciers. The deviation was less than 1% of the total area. In addition, we undertook random evaluations of some glaciers of different size categories. This produced one maximum and one minimum extent by including/excluding each of the ambiguous areas, respectively. The resulting glacier areas appear not to

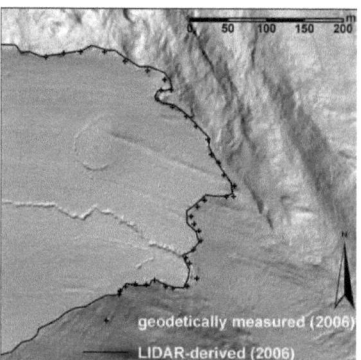

Fig. 11. The glacier margin of Vernagtferner with the LIDAR-derived boundary (solid) and the geodetically measured one (crosses).

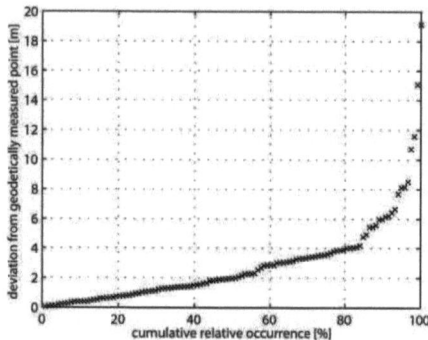

Fig. 12. Absolute deviations of the glacier boundary delineation performed in this study from the geodetically measured margin and their cumulative relative occurrences.

have deviated from each other by more than +/−1.5% of the total glacier-covered area for glaciers bigger than 1 km^2, and up to +/−5% for smaller glaciers.

Further to our application of this essentially qualitative approach that allowed for estimation of a relative error in the overall glacier area, we demonstrated a more quantitative one for all available geodetically measured (combined laser ranger and theodolite) points. They were acquired through the length-change measurement service of the Austrian Alpine Club. Figure 11 shows the delineated glacier boundary of Vernagtferner along with the geodetically measured points (crosses). Both the LIDAR-DEM and the field measurements coincidentally derive from the same date (22 August 2006).

By analogy, we implemented this data as ground truth for 3 more glaciers (Großer Guslarferner, Hochjochferner and Hintereisferner). This resulted in 118 measured points with which to compare the LIDAR interpretation. Figure 12 shows the resultant cumulative relative occurrence of the individual point distances sorted according to their absolute deviation. This can be interpreted as the relative occurrence of measured spots that lie closer to the reference measurement (geodetically) than the respective distance plotted at the y-axes (e.g., 85% of all reference points are within 4 m distance of the margin derived in this paper, or 95% within 8 m). However, while these values are valid for glaciers of different size categories, their impact on the overall area accuracy depends on the size of the glacier itself.

5 Discussion

Figure 13 provides an overview of the spatial resolution and the vertical accuracy of DEMs typically used in glaciology. It also shows the orders of magnitude of the typical mean annual thickness loss of all Austrian glaciers in the last decade (Abermann et al., 2009; Kuhn et al., 2009; and Lambrecht and Kuhn, 2007), of the typical ice thickness loss at debris-free as well as debris-covered parts of Hintereisferner between 2001 and 2005, and of the ice thickness loss at the debris-free margin of Hintereisferner between 1953 and 2003 (Fischer et al., accepted) on the right side of the figure. The remote sensing data outside the rectangular box (orthophotographs and Landsat data) do not include topographic information. References for the individual data sets are given in brackets. The applicability of ice thickness changes for the detection of glacier boundaries depends on the magnitude of elevation change (time difference, climate signal) compared to the root sum square of the vertical accuracies of the DEMs applied. The use of LIDAR-DEMs together with DEM 1997 is thus a comparatively accurate option in terms of both the achieved horizontal resolution and the vertical accuracies.

Compared to other data sets used for glacier boundary delineation, the proposed sets have the advantage of enabling the debris-covered and dead-ice areas to be delineated as well. This had been pointed out, e.g., by Hendriks and Pellikka (2007), as a major disadvantage of multi-spectral methods. Knoll and Kerschner (2009), who worked with LIDAR but did not have multi-temporal DEMs, also addressed this problem.

In order to make use of this advantage, it is worth mentioning the order of magnitude that these surface elevation changes must have in order to be able to definitely distinguish

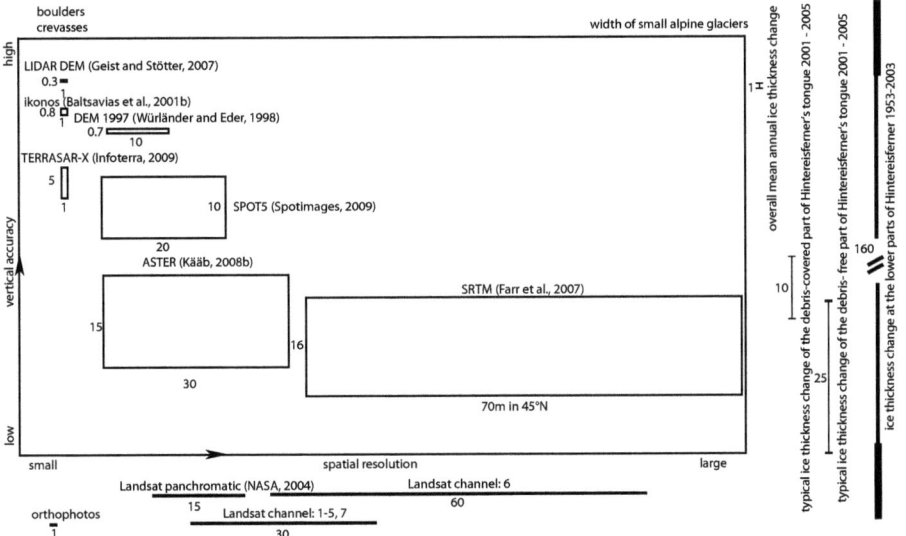

Fig. 13. Schematic distribution of spatial resolution vs. vertical accuracy of commonly used remote sensing data. Orthophotographs and Landsat-scenes do not include vertical information. Typical overall mean annual ice thickness change in Austria's glaciers, values of ice thickness change on debris-free and debris-covered parts of Hintereisferner's glacier margin (2001–2005) along with Hintereisferner's ice thickness loss (1953–2003) is displayed on the right. The lengths and widths of the boxes are scaled, with all numbers in meters. References appear in brackets.

Table 3. Accuracy of the DEM-differences for the examples provided in the paper.

	DEM1	ε_{DEM1} [m]	DEM2	ε_{DEM2} [m]	Δt	$\varepsilon_{\Delta z}$ [m]
Ex. Guslarferner (Fig. 6)	DEM 1997	+/−0.7	LIDAR 2006	+/−0.3	9 yrs	+/−0.8
Ex. Hintereisferner (Fig. 10a)	LIDAR 2001	+/−0.1	LIDAR 2005	+/−0.1	4 yrs	+/−0.1
Ex. Hintereisferner (Fig. 10c)	LIDAR 2004	+/−0.1	LIDAR 2005	+/−0.1	1 yr	+/−0.2
Ex. Reichenkar rock glacier (Fig. 14)	DEM 1997	+/−0.7	LIDAR 2006	+/−0.3	9 yrs	+/−0.8

them from noise introduced by errors in the DEMs. Applying the law of error propagation according to Etzelmüller (2000), the error from the DEM differences ($\varepsilon_{\Delta z}$) is the root of the sum of square errors of the individual DEMs (ε_{DEM1} and ε_{DEM2}),

$$\varepsilon_{\Delta z} = \sqrt{\varepsilon_{DEM1}^2 + \varepsilon_{DEM2}^2}. \tag{1}$$

For the data sets presented in this study, values of $\varepsilon_{\Delta z}$ are summarized in Table 3. For DEM combinations using other data sets $\varepsilon_{\Delta z}$ can be estimated with the vertical accuracies as shown in Fig. 13.

Processes resulting in surface elevation change on ice-free terrain (such as denudation, washout, rockfall or gully formation) could also be misinterpreted as glacier-induced elevation change should they reach the typical orders of magnitude of the latter. Hallet et al. (1996) report "effective rates of glacial erosion" in the Swiss Alps to be around 1 mm/a. Wittmann et al. (2007) quantify "denudation rates" at about 0.9 ± 0.3 mm/a, the same as for the Swiss Alps. Since they are far smaller than typical elevation changes caused by icemelt, these processes can be neglected, even if only elevation changes within one year are considered (e.g., Fig. 9c, cf. typical rates of surface elevation change, Fig. 13). Gully formation along lateral moraines shortly after glacier recession can exceed these values, from typically 23 mm/a (Ballantyne, 2002) to 151 mm/a (Curry et al., 2005) on different Alpine sites. The maximum values found by Schiefer and Gilbert (2007) at 1.5 m/a can be taken as an extreme case of non-glacier-related elevation change. While such formations are visible in a shaded relief of the DEM (Fig. 9b, indicator 1) they do not coincide with strong elevation change, which means that they were formed before 2001. Had they been formed between the DEMs studied, their longitudinal shape would have been visible in the DEM differences. This is one more reason why we applied the method manually, excluding such structures from the resulting glacier area. Rockslides can also cause considerable surface lowering, which might be misinterpreted as glacier area. But they invariably appear with a significant mass gain at the lower section of the rockslide and can thus be distinguished easily.

Another advantage of the glacier boundary delineation with very high resolution DEMs is the ability to implicitly derive volume changes also for very small glaciers, where many other remote sensing data sets fail, simply due to their large spatial resolution or considerable vertical errors (cf. Figs. 5, 6 and 13).

To minimize errors due to interpretation of surface elevation changes introduced by seasonal snow cover we mainly incorporated information from DEM differencing to the ablation areas. In these parts of the glacier fresh snow cover is not significant at the data acquisition dates (late summer or early autumn) compared to the strong elevation changes due to negative mass balances of the last years.

A major disadvantage of the method proposed is the considerable manual digitization effort that is necessary for the derivation of glacier boundaries, requiring between 0.5 and 2 h per glacier, depending on its complexity and size.

6 Conclusions

The inclusion of multi-temporal DEMs with a relative vertical accuracy significantly better than the ice thickness change over the investigated period improves the accuracy of glacier boundary mapping. This method is well suited to study areas with a comparably small extent, where an accurate knowledge of glacier area and volume change is needed, since it requires considerable manual digitization. One great advantage compared with other techniques for glacier boundary delineation is the high degree of accuracy achieved for the delineation of small glaciers (e.g., $< 0.5\,\mathrm{km}^2$). Combining this with additional information such as multi-temporal DEMs and orthophotographs or other remote sensing data further improves the result. A broader application of the developed method was performed by Abermann et al. (2009): a regional update of an existing glacier inventory was undertaken. In addition, a new inventory is planned for other regions in Austria based on this methodology. Geophysical investigations of dead-ice regions could provide further insight into, and better ground truth, regarding these areas. It may well be an interesting topic for future study.

The better the vertical accuracy and the horizontal resolution of the DEMs, the shorter is the time period between the acquisition of the DEMs that is needed to obtain statistically significant elevation changes. When applied to the climatic conditions of glaciers closer to a steady state, this mapping procedure would be less successful as surface elevation changes would also be smaller. The accuracy of the glacier boundary delineation has proven higher in areas with large elevation change, i.e., low elevations and bare ice.

A sequence of multi-temporal airborne LIDAR-DEMs also covering rock glaciers will enhance the importance of this application for conducting studies on detailed elevation changes in creeping permafrost, including the implication of volume changes.

The use of multi-temporal DEMs will be of significant importance for future glaciological applications. The number of accurate high-resolution DEMs is increasing with both airborne and satellite data. The predicted future climate change (Trenberth et al., 2007) will result in continuing glacier volume and area loss. For this reason this method may be extended further and it is planned to make use of the information from high-resolution surface elevation changes to develop a semi-automatic glacier boundary delineation algorithm.

Acknowledgements. This study was funded by the Commission for Geophysical Research, Austrian Academy of Sciences. The LIDAR-DEM 2006 was acquired by the Regional Government of Tyrol. The authors would like to thank M. Kuhn and C. Knoll for their comments, S. Braun-Clarke, E. Dryland and L. Raso, for proofreading the paper as native English speakers, H. Schneider for providing the data for ground truthing and M. Attwenger for providing information on the LIDAR-DEM. Two anonymous referees and the paper's editor, A. Kääb, are gratefully acknowledged for constructive remarks and useful suggestions which improved the manuscript considerably.

Edited by: A. Kääb

References

Abermann, J., Lambrecht, A., Fischer, A., and Kuhn, M.: Quantifying changes and trends in glacier area and volume in the Austrian Ötztal Alps (1969–1997–2006), The Cryosphere, 3, 205–215, 2009,
http://www.the-cryosphere-discuss.net/3/205/2009/.

Andreassen, L. M., Paul, F., Kääb, A., and Hausberg, J. E.: Landsat-derived glacier inventory for Jotunheimen, Norway, and deduced glacier changes since the 1930s, The Cryosphere, 2, 131–145, 2008,
http://www.the-cryosphere-discuss.net/2/131/2008/.

Ballantyne, C. K.: Paraglacial geomorphology, Quaternary Sci. Rev., 21, 1935–2017, 2002.

Baltsavias, E. P., Favey, E., Bauder, A., Bösch, H., and Pateraki, M.: Digital surface modelling by airborne laser scanning and digital photogrammetry for glacier monitoring, The Photogrammetric Record, 17(98), 243–273, 2001a.

Baltsavias, E. P., Pateraki, M., and Zhang, L.: Radiometric and geometric evaluation of IKONOS GEO images and their use for 3D building and modelling, Proceedings of the ISPRS Workshop "High Resolution Mapping from Space 2001", Hannover, Germany, 19–21 September 2001, 2001b.

Curry, A. M., Cleasby, V., and Zukowskyj, P.: Paraglacial response of steep, sediment-mantled slopes to post-"Little Ice Age" glacier recession in the central Swiss Alps, J. Quaternary Sci., 21, 211–225, 2005.

DeBeer, C. M. and Sharp, M. J.: Recent changes in glacier area and volume within the southern Canadian Cordillera, Annals of Glaciology, 46, 215–221, 2007.

Dyurgerov, M. B. and Meier, M. F.: Twenthieth century climate change: Evidence from small glaciers, PNAS, 97(4), 1406–1411, 2000.

Etzelmüller, B.: On the quantification of surface changes using grid-based digital elevation models (DEMs), Transactions in GIS, 4(2), 129–143, 2000.

Farr, T. G., Rosen, P., Caro, E., et al.: The Shuttle Radar Topography Mission, Rev. Geophys., 45, RG2004, doi:10.1029/2005RG000183, 2007.

Favey, E., Wehr, A., Geiger, A., and Kahle, H.-G.: Some examples of European activities in airborne laser techniques and an application in glaciology, J. Geodyn., 34, 347–355, 2002.

Fischer, A.: Glaciers and climate change: Interpretation of 50 years of direct mass balance of Hintereisferner, Global Planet. Change, doi:10.1016/j.gloplacha.2009.11.014, accepted, 2010.

Geist, T., Lutz E., and Stötter, J.: Airborne laser scanning technology and its potential for applications in glaciology, International Archives of Photogrammetry, Remote Sensing and Spatial Information Science, Vol. XXXIV, part 3/W13, 101–106, 2003.

Geist, T. and Stötter, J.: Documentation of glacier surface elevation change with multi-temporal airborne laser scanner data – case study: Hintereisferner and Kesselwandferner, Tyrol, Austria, Zeitschrift für Gletscherkunde und Glazialgeologie, 41, 77–106, 2007.

Google Earth, 47°2'55" N 11° 1'55" E, last access: 18 September 2009.

Haeberli, W., Frauenfelder, R., Hoelzle, M., and Maisch, M.: On rates and acceleration trends of global glacier mass changes, Geogr. Ann. A, 81(4), 585–591, 1999.

Hallet, B., Hunter, L., and Bogen, J.: Rates of erosion and sediment evacuation by glaciers: A review of field data and their implications, Global Planet. Change, 12, 213–235, 1996.

Hendriks, J. P. M. and Pellikka, P.: Semi-automatic glacier delineation from Landsat imagery over Hintereisferner, Zeitschrift für Gletscherkunde und Glazialgeologie, 41, 55–75, 2007.

Höfle, B., Geist, T., Rutzinger, M., and Pfeifer, N.: Glacier surface segmentation using airborne laser scanning point cloud and intensity data, International Archives of Photogrammetry, Remote Sensing and Spatial Information Sciences, Vol. XXXVI/3, 195–200, 2007.

InfoTerra: 3D topographic mapping with TerraSAR-X. Unique mapping concepts using high-resolution spaceborne SAR, 2009.

Kääb, A.: Remote sensing of permafrost – related problems and hazards, Permafrost Periglac., 19, 107–136, 2008a.

Kääb, A.: Glacier volume changes using ASTER satellite stereo and ICESat GLAS laser altimetry. A test study on Edgeoya, Eastern Svalbard, IEEE Transactions on Geoscience and Remote Sensing, 46(10), 2823–2830, 2008b.

Kääb, A., Paul, F., Maisch, M., and Häberli, W.: The new remote-sensing-derived Swiss Glacier Inventory: II. First results, Annals of Glaciology, 34, 362–366, 2002.

Kääb, A. and Vollmer, M.: Surface geometry, thickness changes and flow fields on creeping mountain permafrost: Automatic extraction by digital image analysis, Permafrost Periglac., 11, 315–326, 2000.

Kirkbride, M. P. and Warren, C. R.: Tasman Glacier, New Zealand: 20th-century thinning and predicted calving retreat, Global Planet. Change, 22, 11–28, 1999.

Knoll, C. and Kerschner, H.: A glacier inventory for South Tyrol, Italy, based on airborne laser scanner data, Annals of Glaciology, 50(53), 46–52, 2009.

Kodde, M., Pfeiffer, N., Gorte, B., Geist, T., and Höfle, M.: Automatic glacier surface analysis from airborne laser scanning, IS-PRS Workshop Laser Scanning 2007 XXXVI Part 3/W52, 221–226, 2007.

Krainer, K. and Mostler, W.: Reichenkar Rock Glacier, a glacial derived debris-ice system in the Western Stubai Alps, Austria, Permafrost Periglac., 11, 267–275, 2000.

Krainer, K., Mostler, W., and Span, N.: A glacier derived, ice-cored rock glacier in the western Stubai Alps (Austria): Evidence from exposures and ground penetrating radar investigation, Zeitschrift für Gletscherkunde und Glazialgeologie, 38(Eq. 1), 21–34, 2002.

Krainer, K. and Mostler, W.: Flow velocities of active rock glaciers in the Austrian Alps, Geogr. Ann. A., 88(4), 267–280, 2006.

Kuhn, M.: Zeitschrift für Gletscherkunde und Glazialgeologie, 41, Innsbruck, Austria, 232 pp., 2007.

Kuhn, M., Lambrecht, A., Abermann, J., Patzelt, G., and Gross, G.: Projektbericht 10. Die österreichischen Gletscher 1998 und 1969, Flächen- und Volumenänderungen, Verlag der österreichischen Akademie der Wissenschaften, Wien, Austria, 128 pp., 2009.

Lambrecht, A. and Kuhn, M.: Glacier changes in the Austrian Alps during the last three decades, derived from the new Austrian glacier inventory, Annals of Glaciology, 46, 177–184, 2007.

Lemke, P., Ren, J., Alley, R. J., et al.: Observations: Changes in Snow, Ice and Frozen Ground, in: Climate Change 2007, The Physical Science Basis, Contribution of Working Group I to the Fourth Assessment Report of the Intergovernmental Panel on Climate Change, edited by: Solomon, S., Qin, D., Manning, M., et al., Cambridge University Press, Cambridge, United Kingdom and New York, NY, USA, 2007.

Patzelt, G.: The Austrian glacier inventory: status and first results. Riederalp Workshop 1978 – World Glacier Inventory, IAHS, 1980.

Patzelt, G.: Gletscherbericht 2003/2004, Sammelbericht über die Gletschermessungen des Oesterreichischen Alpenvereins im Jahre 2004, Mitteilungen des Oesterreichischen Alpenvereins, 60(130), 24–31, 2005.

Patzelt, G.: Gletscherbericht 2004/2005, Sammelbericht über die Gletschermessungen des Österreichischen Alpenvereins im Jahre 2005, Bergauf, 2, 6–11, 2006.

Paul, F., Kääb, A., Maisch, M., Kellenberger, T. W., and Häberli,

W.: The new remote-sensing-derived Swiss Glacier Inventory: I. methods, Annals of Glaciology, 34, 355–361, 2002.

Paul, F., Kääb, A. and Haeberli, W.: Recent glacier changes in the Alps observed from satellite: Consequences for future monitoring strategies, Global Planet. Change, 56(1/2), 111–122, 2007.

Paul, F., Barry, R., Cogley, G., Frey, H., Haeberli, W., Ohmura, A., Ommanney, S., Raup, B., Rivera, A., and Zemp, M.: Recommendations for the compilation of glacier inventory data from digital sources, Annals of Glaciology, 50(53), 119–126, 2009.

Pellikka, P. and Rees, W. G.: Remote sensing of glaciers – Techniques for topographic, spatial and thematic mapping of glaciers, London, UK, 340 pp., 2009.

Rott, H. and Markl, G.: Improved snow and glacier monitoring by the Landsat Thematic Mapper, Proceedings of a workshop on Landsat Thematic Mapper applications, ESA, SP-1102, 3–12, 1999.

Schiefer, E. and Gilbert, R.: Reconstructing morphometric change in a proglacial landscape using historical aerial photography and automated DEM generation, Geomorphology, 88, 167–178, 2007.

Spotimages: http://www.spotimage.com/automne_modules_files/standard/public/p811_9d709b1bd850b040110d9d66db425dd2SPOT_DEM_EN_140509.pdf, last access: 20 October 2009.

Trenberth, K. E., Jones, P.D., Adler, R., et al.: Observations: Surface and Atmospheric Climate Change. In: Climate Change 2007: The Physical Science Basis, Contribution of Working Group I to the Fourth Assessment Report of the Intergovernmental Panel on Climate Change, Cambridge University Press, Cambridge, United Kingdom and New York, NY, USA, 2007.

Tirismaps: https://portal.tirol.gv.at, last access: 29 May 2009.

UNESCO: Perennial ice and snow masses: a guide for compilation and assemblage of data for a world inventory, Technical Paper Hydrology 1. UNESCO/IASH, 1970.

Wittmann, H., von Blanckenburg, F., Kruesmann, T., Norton, K. P., and Kubik, P. W.: Relation between rock uplift and denudation from cosmogenic nuclides in river sediment in the Central Alps of Switzerland, J. Geophys. Res., 112, F04010, doi:10.1029/2006JF000729, 2007.

Würländer, R. and Eder, K.: Leistungsfähigkeit aktueller photogrammetrischer Auswertemethoden zum Aufbau eines digitalen Gletscherkatasters, Zeitschrift für Gletscherkunde und Glazialgeologie, 35, 167–185, 1998.

Würländer, R., Eder, K., and Geist, T.: High quality DEMs for glacier monitoring – image matching versus laser scanning, International Archives of Photogrammetry, Remote Sensing and Spatial Information Science, XXXV, Part B7, 2004.

A3: PAPER III: QUANTIFYING CHANGES AND TRENDS IN GLACIER AREA AND VOLUME IN THE AUSTRIAN ÖTZTAL ALPS (1969-1997-2006)

By: Abermann, J., A. Lambrecht, A. Fischer and M. Kuhn. *Published in The Cryosphere*, **3**(2): 205-215.

Quantifying changes and trends in glacier area and volume in the Austrian Ötztal Alps (1969-1997-2006)

J. Abermann[1,2], **A. Lambrecht**[2], **A. Fischer**[2], **and M. Kuhn**[1,2]

[1]Austrian Academy of Sciences, Commission for Geophysical Research, Vienna, Austria
[2]Institute of Meteorology and Geophysics, University of Innsbruck, Innsbruck, Austria

Received: 9 June 2009 – Published in The Cryosphere Discuss.: 1 July 2009
Revised: 24 September 2009 – Accepted: 28 September 2009 – Published: 20 October 2009

Abstract. In this study we apply a simple and reliable method to derive recent changes in glacier area and volume by taking advantage of high resolution LIDAR (light detection and ranging) DEMs (digital elevation models) from the year 2006. Together with two existing glacier inventories (1969 and 1997) the new dataset enables us to quantify area and volume changes over the past 37 years at three dates. This has been done for 81 glaciers (116 km^2) in the Ötztal Alps which accounts for almost one third of Austria's glacier extent. Glacier area and volume have reduced drastically with significant differences within the individual size classes. Between 1997 and 2006 an overall area loss of 10.5 km^2 or 8.2% occurred. Volume has reduced by 1.0 km^3 which accounts for a mean thickness change of −8.2 m. The availability of three comparable inventories allows a comprehensive size and altitude dependent analysis of glacier changes but lacks a high temporal resolution. For the comparison of rates of changes between the two different periods (1969 to 1997 with 1997 to 2006) we propose two approaches in this study: a) to estimate mean overall rates of changes (including a period of advance) and b) to extract periods of net-retreat by using additional information (length change and mass balance measurements). Analysis of the resulting acceleration factors reveals that the retreat of volume and mean thickness changes has accelerated significantly more than that of area changes.

1 Introduction

For the past decades a general glacier mass and area loss has been observed all over the world with few exceptions (e.g. Lemke et al., 2007; Oerlemans, 2005; Dyurgerov and

Correspondence to: J. Abermann
(jakob.abermann@uibk.ac.at)

Meier, 2000; Haeberli, 1999) and especially in the Alps (Lambrecht and Kuhn, 2007; Citterio et al., 2007; Kääb et al., 2002). In order to document the glacier extent, various national glacier inventories have been produced using different remote sensing techniques, such as photogrammetry (e.g. Kuhn et al., 2009; Lambrecht and Kuhn, 2007; Schneider et al., 2007), satellite data (e.g. Paul et al., 2002a; Andreassen et al., 2008) or LIDAR (e.g. Knoll and Kerschner, 2009). In Austria, two complete glacier inventories exist on the basis of analogue (1969) and digital (1998) airborne photogrammetry, respectively. The first one (1969) was compiled by Patzelt (1980) and Groß (1987) and later digitised during the compilation of the second Austrian inventory (data acquisition: 1996 to 2002). The data acquisition year of the second inventory for the Ötztal Alps was 1997 (Kuhn et al., 2009; Lambrecht and Kuhn, 2007). Paul (2002b) investigated glacier changes of sub-regions of Tyrolean glaciers with Landsat data. Detailed maps of selected glaciers in the study area were produced between 1969 and 1997 (e.g. Hintereisferner: Kuhn, 1979, or Vernagtferner: Heipke et al., 1994; Endres, 2001).

The strong area and volume loss of the last decade raised interest in an updated glacier inventory. Based on LIDAR-DEMs this was developed for the glaciologically well investigated southern part of the Austrian Ötztal Alps and is presented in this study. Besides forming an important source for the determination of glacier extent as its main purpose, periodically updated glacier boundaries are a necessary input for accurate mass balance studies (e.g. Elsberg et al., 2001; Fischer et al., 2009) and modelling of the future glacier extent. A sequence of glacier inventories allows for the quantification of the accelerated glacier change in the context of a changing climate.

Our first aim in this study is thus to present area and volume changes that occurred in the Ötztal Alps during the last decades, derived from LIDAR-DEMs in combination with former inventories. We also include size as well as altitude

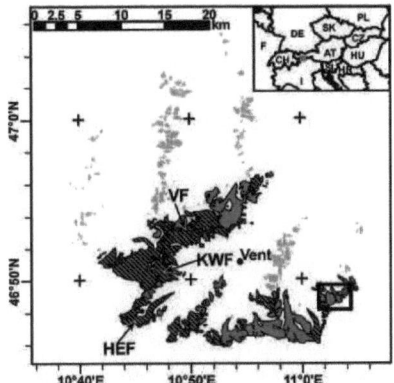

Fig. 1. Study area Ötztal Alps (grey: glacier extent 1997 of the glaciers in the Ötztal Alps that have not been updated in our study, red: updated glacier extent 2006, dashed: glaciers with length measurement records in the study area, black rectangle: extent of Fig. 6). HEF, KWF and VF refer to three glaciers with mass balance records (Hintereisferner, Kesselwandferner and Vernagtferner). The valley station Vent, where the climate data from Fig. 2 derives from, is indicated by a black circle.

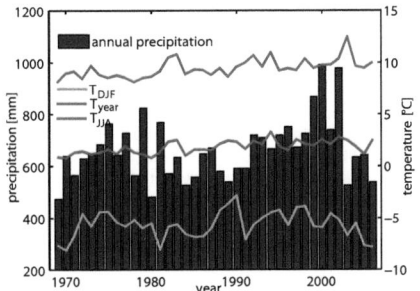

Fig. 2. Annual precipitation sums (blue bars) and winter – (December, January, February; grey), annual mean – (green) and summer air temperature evolution (June, July, August; red) for the period 1969–2006.

dependent trend considerations. Secondly, we derive overall acceleration trends of glacier changes by comparing mean annual area and volume changes between the two periods investigated (1969–1997 and 1997–2006). To estimate the acceleration of retreat we propose the use of additional information to extract periods of net-retreat (length change and mass balance measurements).

2 Study area

The Ötztal Alps are located in a central-alpine dry region at around 47° N and 11° E (Fliri, 1975). Their Austrian part is situated on the northern slope of the main alpine divide. Figure 1 shows the study area with the glacier extent of all Austrian Ötztal glaciers in 1997 (grey) and all updated glaciers in 2006 (red). In total, 81 glaciers of all size and aspect classes were remeasured, chosen simply by the extent of the DEM used and covering all well studied glaciers in this area. This new inventory accounts for 84% of the glacier area in this mountain range (1997: 151 km^2) and represents approx. 27% of the whole Austrian glacier area (1998: 470 km^2).

Figure 2 shows the climatic conditions of the study area at the example of the valley station Vent at 1900 m a.s.l. (data from the Institute of Meteorology and Geophysics, University of Innsbruck). In the period 1997–2006, annual precipitation showed an interannual variation between ca. 500 mm and 1000 mm with a mean of 660 mm and a standard variation of 119 mm. A sequence of positive precipitation anomalies occurred in the mid-1970s followed by relatively average 1980s and 1990s. Around 2000 there were some strongly positive precipitation anomalies followed by drier years since 2003.

Winter temperatures (DJF) show a larger inter-annual variability (standard deviation: 1.3°C) than mean annual (standard deviation: 0.6°C) and summer temperatures (JJA, (standard deviation: 0.9°C) but no significant trend. A temperature increase was observed in the summer temperatures: the extraordinarily hot summer of 2003 marked the maximum of the period studied. The annual average temperature during the 37 years is 1.7°C.

3 Data

3.1 Glacier inventory data

Two existing inventories were used for the determination of area and volume changes of the past four decades. The first one was established using aerial photographs from the year 1969 (Patzelt, 1980; Groß, 1987) and was digitized later on (Lambrecht and Kuhn, 2007). In the years 1996–2002 a new inventory was produced by means of digital photogrammetry. The results for all Austrian glaciers were then temporally homogenized for the year 1998 (Kuhn et al., 2009; Lambrecht and Kuhn, 2007). The DEMs of the Ötztal Alps for the second inventory have been acquired on the basis of orthophotos (acquisition date: 11 September 1997), therefore we refer to this date further on. The vertical accuracy of these DEMs in general is better than ±1.9 m, according to Lambrecht and Kuhn (2007) and better than ±0.71 m according to Würlander and Eder (1998). Only in extreme cases can errors exceed this value (Abermann et al., 2007).

Several studies show that LIDAR or ALS (airborne laser scanning) is a powerful tool to generate DEMs from glacier covered as well as glacier-free areas (e.g. Kennett and Eiken, 1997; Baltsavias et al., 2001; Geist et al., 2005; Geist and Stötter, 2007; Kodde et al., 2007). The DEMs of 2006 were acquired by high resolution airborne LIDAR undertaken by the Regional Government of Tyrol between 23 August 2006 and 9 September 2006 with vertical errors of ±0.1 m. The acquisition dates make them closely comparable with the data of 11 September 1997 both being around the date of minimum snow extent. Details about the technical specifications of the DEMs used including general remarks on glacier boundary delineation from DEMs are summarized in Abermann et al. (2009).

3.2 Glacier length change measurements

Variations of glacier lengths are determined annually by the length change measurement service of the Austrian Alpine club for a large number of glaciers with the results published annually in the club's magazine (until 2003/2004: e.g. Patzelt, 2005; from 2004/2005: e.g. Patzelt, 2006). Within the study area length change measurements have been performed annually on 16 glaciers since 1969. Glacier names, size and location of these glaciers are summarized in Table 1.

3.3 Mass balance measurements

To estimate annual averages of volume and mean thickness changes between the inventories we used the temporal course of the available mass balance series. On three glaciers within the study area direct mass balance measurements are performed annually. They are performed by the Institute of Meteorology and Geophysics, University of Innsbruck, on Hintereis- and Kesselwandferner (Fischer and Markl, 2009; Kuhn et al., 1999). On Vernagtferner the Commission for Glaciology, Bavarian Academy of Sciences and Humanities, performed the measurements (Glaciology, 2009[1]).

4 Methods

4.1 Area and volume changes

Abermann et al. (2009) proposed a method to delineate glacier boundaries by using hillshades of high-resolution DEMs as a primary data source and including information of multi-temporal DEMs as well as orthophotos to increase accuracies in ambiguous areas. Applying this method, we updated the glacier boundaries of 2006 on the basis of the inventory from 1997.

[1] Available at: http://www.glaciology.de/.

Table 1. Name, area and coordinates of the glaciers for which annual length change measurements are available for the whole investigation period (1969–2006).

Name	area 2006 [km^2]	lon [° E]	lat [° N]
Diemferner	2.34	11.07	46.83
Gaißbergferner	1.03	11.02	46.78
Gepatschferner	16.62	10.95	46.81
Großer Guslarferner	1.40	10.91	46.79
Hintereisferner	7.49	10.86	46.77
Hochjochferner	6.07	10.82	46.79
Kesselwandferner	3.82	10.76	46.80
Langtalerferner	2.62	10.79	46.84
Mutmalferner	0.56	10.80	46.85
Niederjochferner	1.87	10.82	46.88
Rettenbachferner	1.48	10.88	46.88
Rofenkarferner	1.14	10.93	46.93
Sexegertenferner	1.96	10.76	46.85
Taschachferner E	5.71	10.71	46.85
Vernagtferner	8.32	10.85	46.90
Weißseeferner	2.59	10.80	46.89

We kept ice divides constant during the delineation since they have not significantly changed over the past 40 years compared to the strong changes in the ablation area and thus played a somewhat negligible role in our considerations. Furthermore, we address overall changes in ice cover rather than changes of individual glaciers, therefore a shift of an ice divide does not change the results. We also remained consistent with the former inventories by including dead ice bodies and debris-covered areas to the glacier area. If it was not possible to decide whether adjoining snow and firn areas cover ice or rocks, we included them to the glacier area as proposed in UNESCO (1970) and Paul et al. (2009). The resulting glacier masks were identified according to the nomenclature of 1969 which is based on a systematic numbering within drainage areas. Thus, glaciers have not been relabelled, even if they have disintegrated since the last inventory.

For the calculation of volume changes we first resampled all DEMs to the same cell size (5×5 m). This is done to avoid errors introduced by interpolation. Then we subtracted one DEM from the other and multiplied the elevation difference per cell with the cell size (5×5 m) to obtain volume changes.

The calculation of the mean thickness change has been performed by dividing the total volume change by a mean area for the respective period, which gives

$$\Delta \bar{z}_{1969-1997} = \frac{\Delta V_{1969-1997}}{0.5 \cdot (A_{1969} + A_{1997})} \quad (1)$$

for the period 1969–1997, or

$$\Delta \bar{z}_{1997-2006} = \frac{\Delta V_{1997-2006}}{0.5 \cdot (A_{1997} + A_{2006})} \quad (2)$$

for the period 1997–2006.

Abermann et al. (2009) estimated the standard error of the applied method to be ±1.5% of the total area for glaciers larger than 1 km² and up to ±5% for smaller glaciers. Ground truthing of selected glaciers reveals absolute horizontal deviations of less than 4 m for 80% of the sample (Abermann et al., 2009).

4.2 Estimating acceleration trends

It is challenging to compare different periods with differing glacier changes. Whereas the first period contains a significant glacier advance, followed by a strong retreat, the second mainly consists of strong negative area and volume changes. We propose two ways of drawing a comparison between the two periods: in Sect. 4.2.1. we derive overall acceleration factors on the basis of the whole investigation periods, in Sect. 4.2.2. we undertake a temporal reduction to obtain acceleration factors of periods of net-retreat only.

4.2.1 Overall acceleration factors: area, volume and mean thickness changes

As a first step we introduce overall rates of changes where we divide the overall area/volume loss by the overall time period between the inventories for both periods (e.g. Δt_{69-97}=28 years, Δt_{97-06}=9 years). Out of these rates of changes we are able to derive the following acceleration factors:

$$F_{A\_overall} = \frac{\frac{\Delta A_{97-06}}{\Delta t_{97-06\_overall}}}{\frac{\Delta A_{69-97}}{\Delta t_{69-97\_overall}}}, \quad (3)$$

$$F_{A\%\_overall} = \frac{\frac{\Delta A\%_{97-06}}{\Delta t_{97-06\_overall}}}{\frac{\Delta A\%_{69-97}}{\Delta t_{69-97\_overall}}}, \quad (4)$$

$$F_{V\_overall} = \frac{\frac{\Delta V_{97-06}}{\Delta t_{97-06\_overall}}}{\frac{\Delta V_{69-97}}{\Delta t_{69-97\_overall}}}, \quad (5)$$

$$F_{\bar{z}\_overall} = \frac{\frac{\Delta \bar{z}_{97-06}}{\Delta t_{97-06\_overall}}}{\frac{\Delta \bar{z}_{69-97}}{\Delta t_{69-97\_overall}}}. \quad (6)$$

4.2.2 Temporally reduced acceleration factors:

Area changes

Area and volume reduction of the first period has occurred only during a part of the overall time period due to the significant glacier advance around 1980 (Patzelt, 1985). Therefore we decided to perform a temporal reduction on the basis of length change measurements for acceleration factors related to area changes (F_{A_retr}, $F_{A\%\_retr}$) and of mass balance measurements for acceleration factors related to volume changes

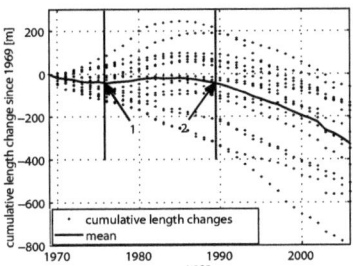

Fig. 3. Cumulative length change for all glaciers that cover the entire period (crosses) and the arithmetic mean over these curves (solid). Arrow 1 indicates the derived onset of the advance period (1976), arrow 2 the state where the same cumulative length as in 1 has been reached again (1989).

(F_{V_retr}, $F_{\bar{z}\_retr}$). In contrast to the above factors (Eqs. 3 to 6) that we interpret as describing altered glacier evolution, the temporally reduced factors are intended to reveal information on accelerated glacier retreat, as described below.

All length change measurements within the study area which cover the whole investigation period (16 glaciers, for details see Table 1) are plotted in Fig. 3. The solid line shows their arithmetic mean. Individual glaciers respond very differently to climatic condition change: some, which have responded with a decreased rate of negative length change, have shown continuous retreat since 1969; others, which have advanced significantly, had not reached their initial state before 2003. Kuhn et al. (1985), using two glaciers within the study area as examples, examined the different glacier behaviour that results from different glacier characteristics (e.g. area-elevation distribution, vertical balance profile or aspect). As the sign of length change correlates well with the sign of area change, we used the temporal course of the arithmetic mean to define the mean onset of the general glacier advance (1976, indicator 1 in Fig. 3). Furthermore, we defined the year in which the same length as recorded at the beginning of the advance was reached again (1989, indicator 2). The observed length change between 1969 and 1997 thus occurred within 15 years in two phases (1969–1976 and 1989–1997). To estimate the mean annual area loss in the period between the first two inventories ($\Delta A/\Delta t_{69-97\_retr}$) we therefore divided the absolute area changes by 15 which is the number of years with net length (and probably area) reduction.

For the period 1997–2006 a continuous length reduction occurred. Therefore we derived estimates for mean annual area change by dividing the overall changes by 9 ($\Delta A/\Delta t_{97-06\_retr}$).

Changes in length and area are not to be compared between two different periods directly since they do not refer to the same absolute length/area. For this reason we assess relative area changes by subtracting the average annual area change value ($\Delta A/\Delta t_{69-97\_retr}$ and $\Delta A/\Delta t_{97-06\_retr}$) from the initial glacier area for each time-step (year). In this way we can calculate a mean annual percental area change, referring to a shrunken absolute value for each year. This has been performed with all individual size classes during both investigated periods again considering the same periods of net-area loss and results in $\Delta A\%/\Delta t_{97-06\_retr}$ and $\Delta A\%/\Delta t_{69-97\_retr}$.

Out of these rates of net-retreat we were now able to calculate F_{A_retr} and $F_{A\%\_retr}$, analogously to Eqs. (3) and (4) but with a reduced Δt for the period 1969–1997 (Δt_{97-06_retr}=15 years).

Volume and mean thickness changes

Figure 4 shows the temporal evolution of the cumulative mean specific mass balance of three glaciers within the study area and their arithmetic mean. To estimate average annual volume and mean thickness change values we extracted the year of the onset of mean mass gain (1973, indicator 3 in Fig. 4). The gain in mass during the positive mass balance years had been lost by 1985, after several years of negative mass balances (indicator 4), which leads to a Δt_{97-06_retr} of 16 years (16 out of 28 years contributed to the observed volume loss). For the second period (1997–2006), one year (2001) showed a positive sum of specific mass balance which leads to a Δt_{97-06_retr} of 8 years and analogously to Eq. (5) we calculated F_{V_retr}. Since the mean thickness change is directly connected with the volume loss (Eq. 1), we used the same lengths of intervals to derive mean annual values of $\Delta z/\Delta t_{69-97\_retr}$ and $\Delta z/\Delta t_{97-06\_retr}$, and consequently $F_{\bar{z}\_retr}$.

5 Results

Glacier area, all other glacier inventory attributes, absolute and relative area changes, as well as absolute volume and mean thickness changes were determined and calculated for each individual glacier within the study area for the year 2006 (81 glaciers, 116 km²). Three out of originally 84 glaciers (1997) in the study area shrank in size from the smallest class in 1997 (between 0.01 km² and 0.1 km²) below the minimum size of 0.01 km² in 2006 to be accounted for as a glacier (UNESCO, 1970; Paul et al., 2009).

Figure 5a shows the area-altitude distribution of all updated glaciers for 1969, 1997 and 2006. Areas with the largest ice cover are between 3100 and 3200 m a.s.l. The minimum altitude of ice cover rose from 2060 m a.s.l. to 2120 m a.s.l. between 1969 and 2006. It is worth noting that the elevation of maximum ice-cover for the study area, as shown in Fig. 5a, is about 200 m higher (approx. 3200 m) than of all Austrian glaciers (Lambrecht and Kuhn, 2007, Fig. 3). This is due to the central-alpine dry climate of the

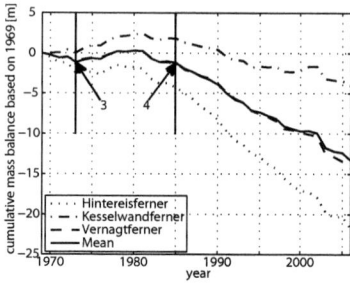

Fig. 4. Cumulative mean specific mass balance of the glaciers Hintereisferner, Kesselwandferner and Vernagtferner and their mean refering to 1969. It is pure coincidence that the cumulative balance values of the mean and of Vernagtferner follow almost exactly the same temporal course. Arrow 3 indicates the year when an overall mass gain starts (1973) and by the year 1985 (arrow 4), the same volume was reached again.

Ötztal Alps as well as the relatively large areas that are available for potential glaciation in these high altitudes.

The absolute area changes (Fig. 5b) show that for both periods the strongest area loss occurred at about 3000 m a.s.l. There are, however, some significant differences considering the distribution over the entire altitude range. Up to approximately 2900 m a.s.l. both periods show similar values of absolute area changes, while in higher elevations the area reduction is about one third weaker for the period 1997–2006 than for 1969–1997. This means that for lower elevations the absolute area change within the last decade is similar to that of the 28 years before.

Figure 5c shows the vertical distribution of relative area changes that occurred in the respective period. The solid line (1969–1997) refers to the absolute value of 1969, the dashed line to the absolute value of 1997. Thus, the same relative change (%-value) reflects a larger absolute change for the second period because the reference value is smaller already. The lowest elevation band has disappeared (−100%), and up to an elevation of about 3000 m relative changes in the second period are more negative than in the first period with a reversing sign above. Above 3650 m positive relative area changes occur, although this fact should not be overinterpreted since it represents only a very small area (about 1 ha at elevations higher than 3650 m).

Figure 6a shows an example of area changes for Rotmoos- and Wasserfallferner in the southern Ötztal Alps. The glacier boundaries of the three dates (1969, 1997 and 2006) are plotted over the hillshade. Large rock outcrops developed on Rotmoosferner between 1997 and 2006. In 2005 the two glaciers separated, a phenomenon that is observed on various alpine glaciers during the glacier retreat of the past decades

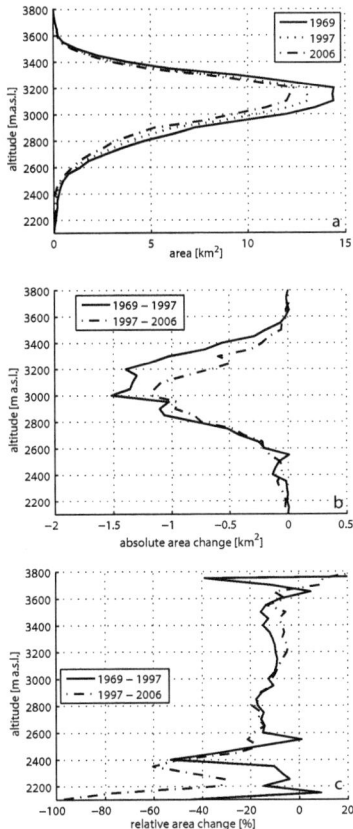

Fig. 5. Area-altitude distribution in 50 m-intervals of all updated glaciers for 1969, 1997 and 2006 **(a)** and absolute area changes of the 50 m-intervals for the periods 1969–1997 and 1997–2006 **(b)**. Figure 5c shows the vertical distribution of relative area changes referring to 1969 (solid) and to 1997 (dashed).

(e.g. Knoll and Kerschner, 2009; Paul et al., 2004). Elevation changes between 1997 and 2006 are plotted in Fig. 6b. An interesting detail is the slightly positive thickness change on Wasserfallferner, indicating the contrast of a glacier with a "healthy" accumulation zone (WFF) to a valley glacier with a relatively thick glacier tongue (RMF). Such a behaviour is also supported by other studies (e.g. Paul and Haeberli, 2008; Pelto, 2006). One possible reason could be that the positive precipitation anomalies of the early 2000s (Fig. 2) have caused this slight gain in surface elevation of this relatively high situated glacier.

The quantitative results for area, volume and mean thickness changes are summarized in Table 2. For this purpose the glaciers are divided into size classes based on their area in 2006. An overall absolute area change of $-17.6\,\mathrm{km}^2$ between 1969 and 1997 and $-10.5\,\mathrm{km}^2$ for 1997–2006 has been derived. This corresponds to a relative area change of -12.2% for 1969–1997 (referring to 1969) and -8.3% for 1997–2006 (referring to 1997). The respective values for the volume changes for these periods are $-1.3\,\mathrm{km}^3$ and $-1.0\,\mathrm{km}^3$ and $-9.5\,\mathrm{m}$ and $-8.2\,\mathrm{m}$ for the mean thickness changes. In general our results are similar but slightly more negative than the values that are derived by Knoll and Kerschner (2009) for the South Tyrolean glaciers (mean thickness change 1997–2006: $-7.0\,\mathrm{m}$). In the period 1969–1997 the relative area loss decreases with increasing glacier size as plotted in Fig. 7. The largest values are observed for the smallest class ($<0.1\,\mathrm{km}^2$, -52.2%). The second period shows the strongest relative area losses for glaciers that are between 0.1 and $0.5\,\mathrm{km}^2$ in area (-16.5%). Very small glaciers ($<0.1\,\mathrm{km}^2$) still lost a considerable fraction of their area in the second period (-11.8%) but the rates of changes have decelerated compared to the first period, where 52.2% of the overall area were lost. Changes in this size class in neighbouring South Tyrol are slightly less negative (-7.7% between 1997 and 2006 according to Knoll and Kerschner, 2009).

The two methods of comparing the rates of glacier changes of the two periods are quantitatively summarized in Tables 3 and 4. A direct comparison of the acceleration factors as derived by two different methods is plotted in Fig. 8a and b. The bars represent the respective factors and their distribution throughout the size-classes. These factors allow for investigation of glacier area and volume evolution within size classes and permit trends in absolute area change with relative, volume and mean thickness changes to be contrasted. An acceleration factor larger than 1 means that mean annual changes have increased, comparing the second period with the first period. Considering the overall periods (including the period of advance), acceleration factors vary between 0.34 (F_{A_overall} for the smallest class) and 3.94 ($F_{\overline{z}\_\mathrm{overall}}$ for glaciers between $0.5\,\mathrm{km}^2$ and $1\,\mathrm{km}^2$). Acceleration factors for the overall periods and the sum of all size classes lie between 1.86 for absolute area (F_{A_overall}) and 2.67 ($F_{\overline{z}\_\mathrm{overall}}$) for mean thickness changes.

For a quantitative interpretation of the individual acceleration factors we focus on the ones derived for periods of net-retreat in the following. The general distribution over the size classes however, is very similar for both applied methods, with the major difference being their differing absolute values (cf. Fig. 8a and b).

F_{A_retr} shows that the mean absolute annual area changes remained the same during the periods of retreat 1997 to 2006 compared with 1969 to 1997. Only in the three largest classes

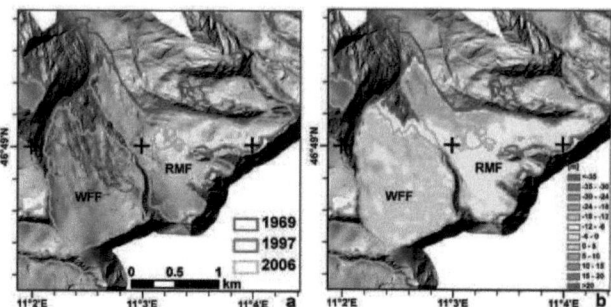

Fig. 6. Rotmoosferner (RMF), Wasserfallferner (WFF) and their surrounding glaciers: Hillshade and glacier boundary 1969, 1997 and 2006 (**a**); glacier boundaries as well as thickness change 1997–2006 (**b**).

Table 2. Summary of glacier covered area of each size class (1969, 1997 and 2006), absolute and relative area changes, volume and mean thickness changes for the periods (1969–1997 and 1997–2006).

Class [km^2]	Count 06	A 69 [km^2]	A 97 [km^2]	A 06 [km^2]	ΔA 69–97 [km^2]	%	ΔA 97–06 [km^2]	%	ΔV 69–97 [*10^6 m^3]	ΔV 97–06 [*10^6 m^3]	Δ$\bar{z}_{69-97}$ [m]	Δ$\bar{z}_{97-06}$ [m]
10–20	1	18.0	17.2	16.6	−0.8	−4.4	−0.5	−3.1	−129	−78	−7.4	−4.6
5–10	7	64.5	57.8	53.2	−6.7	−10.4	−4.6	−8.0	−688	−554	−11.2	−10.0
1–5	15	38.7	34.3	31.3	−4.4	−11.4	−3.0	−8.7	−331	−239	−9.1	−7.3
0.5–1	11	10.1	8.2	7.3	−1.8	−18.2	−0.9	−10.9	−57	−61	−6.2	−7.8
0.1–0.5	25	10.1	7.8	6.5	−2.4	−23.3	−1.3	−16.5	−63	−52	−7.0	−7.4
0.01–0.1	22	2.8	1.4	1.2	−1.5	−52.2	−0.2	−11.8	−19	−7	−9.3	−5.1
All	81	144.2	126.6	116.1	−17.6	−12.2	−10.5	−8.3	−1286	−990	−9.5	−8.2

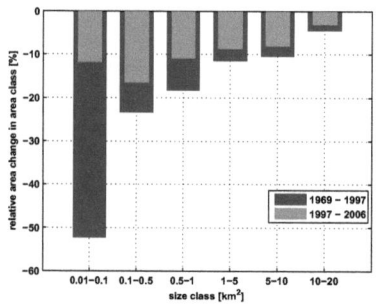

Fig. 7. Relative area changes for the period 1969–1997 (blue) and 1997–2006 (orange) for the individual size classes.

(>5 km^2) an acceleration of 13 to 15% (F_{A_retr}=1.13 to 1.15) occurred. This means that for large glaciers the absolute area loss is slightly stronger in the time period 1997 to 2006 than for the 15 years with net area reduction (within a total of 28 years) before. The very small glaciers have decelerated significantly in terms of mean absolute annual area changes (F_{A_retr}=0.18). We point out that a direct comparison of absolute values of area changes is doubtful. Their relative impact on overall glacier changes depends on the absolute glacier extent they refer to, which may have changed between the two periods.

Therefore we propose the use of relative area changes ($F_{A\%\_retr}$) in order to enable a comparison between time periods and also between size classes. In contrast to absolute area changes (F_{A_retr}=1.00) that did not accelerate, the mean annual relative area change increased by about 10% between the two periods ($F_{A\%\_retr}$=1.11) due to the reduced remaining area. The lowest values of $F_{A\%\_retr}$ are observed for the smallest glaciers ($F_{A\%\_retr}$=0.29). The strongest increase in relative annual glacier area loss ($F_{A\%\_retr}$=1.27) occurs for the glaciers between 5 and 10 km^2 in size.

Table 3. Mean rates of changes referring to the overall periods and referring to reduced periods of net-retreat for absolute and relative area changes. Note, that the rates of changes are displayed in ha/a whereas absolute areas (Table 2) are given in km².

Class	$\Delta A/\Delta t$ 6997_overall	$\Delta A/\Delta t$ 9706_overall	$F_{A\_overall}$	$\Delta A/\Delta t$ 6997_retr	$\Delta A/\Delta t$ 9706_retr	F_{A_retr}	$\Delta A\%/\Delta t$ 6997_overall	$\Delta A\%/\Delta t$ 9706_overall	$F_{A\%\_overall}$	$\Delta A\%/\Delta t$ 6997_retr	$\Delta A\%/\Delta t$ 9706_retr	$F_{A\%\_retr}$
[km²]	[ha/a]	[ha/a]		[ha/a]	[ha/a]		[%/a]	[%/a]		[%/a]	[%/a]	
10–20	−2.8	−6.0	2.10	−5.3	−6.0	1.13	0.16	0.35	2.26	−0.30	−0.35	1.17
5–10	−23.9	−51.6	2.15	−44.7	−51.6	1.15	0.38	0.93	2.45	−0.73	−0.93	1.27
1–5	−15.8	−33.3	2.11	−29.4	−33.3	1.13	0.42	1.01	2.43	−0.80	−1.01	1.26
0.5–1	−6.6	−10.0	1.53	−12.2	−10.0	0.82	0.69	1.28	1.85	−1.33	−1.28	0.96
0.1–0.5	−8.4	−14.2	1.68	−15.7	−14.2	0.90	0.91	1.98	2.17	−1.76	−1.98	1.13
0.01–0.1	−5.3	−1.8	0.34	−9.8	−1.8	0.18	2.51	1.39	0.55	−4.79	−1.39	0.29
All	−62.8	−116.8	1.86	−117.2	−116.8	1.00	0.45	0.96	2.14	−0.86	−0.96	1.11

Table 4. Mean rates of changes referring to the overall periods and referring to reduced periods of net-retreat for absolute volume and mean thickness changes.

Class	$\Delta V/\Delta t$ 69–97_overall	$\Delta V/\Delta t$ 97–06_overall	$F_{V\_overall}$	$\Delta V/\Delta t$ 69–97_retr	$\Delta V/\Delta t$ 97–06_retr	F_{V_retr}	$\Delta z/\Delta t$ 69–97_overall	$\Delta z/\Delta t$ 97–06_overall	$F_{\bar{z}\_overall}$	$\Delta z/\Delta t$ 69–97_retr	$\Delta z/\Delta t$ 97–06_retr	$F_{\bar{z}\_retr}$
[km²]	[*10⁶ m³/a]	[*10⁶ m³/a]		[*10⁶ m³/a]	[*10⁶ m³/a]		[m/a]	[m/a]		[m/a]	[m/a]	
10–20	−5	−9	1.87	−8	−10	1.20	−0.3	−0.5	1.95	−0.5	−0.6	1.25
5–10	−25	−62	2.51	−43	−69	1.61	−0.4	−1.1	2.76	−0.8	−1.3	1.78
1–5	−12	−27	2.24	−21	−30	1.44	−0.3	−0.8	2.50	−0.6	−0.9	1.60
0.5–1	−2	−7	3.35	−4	−8	2.16	−0.2	−0.9	3.94	−0.4	−1.0	2.53
0.1–0.5	−2	−6	2.61	−4	−7	1.68	−0.2	−0.8	3.27	−0.5	−0.9	2.10
0.01–0.1	−1	−1	1.04	−1	−1	0.67	−0.3	−0.6	1.70	−0.6	−0.6	1.09
All	−46	−109	2.38	−80	−124	1.54	−0.3	−0.9	2.67	−0.6	−1.0	1.72

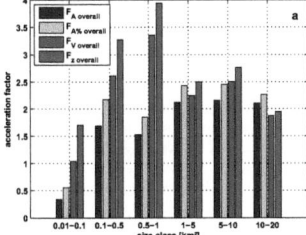 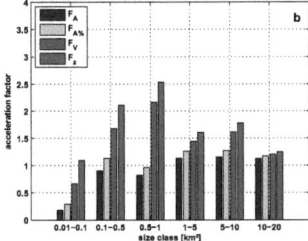

Fig. 8. Acceleration factors referring to mean annual area (absolute and relative) as well as volume and mean thickness changes. In (**a**), no temporal reduction has been performed, the overall area change has been divided by the time between the inventories. In (**b**) periods of net-retreat are extracted as explained in Sect. 4.2.

The mean annual ice volume losses have increased by more than 50% (F_{V_retr}=1.54), while the mean annual thickness reductions by more than 70% ($F_{\bar{z}\_retr}$=1.72). Analysing the different size classes reveals that the strongest changes occurred for glaciers between 0.5 and 1 km², where the mean annual volume change as well as the mean annual thickness change more than doubled (F_{V_retr}=2.16, $F_{\bar{z}\_retr}$=2.53) compared to the period of net-retreating within 1969–1997. Again, glaciers of the smallest size class show a decrease in mean annual volume changes compared to the earlier period (F_{V_retr}=0.67). This is in line with trends in area changes that have been highlighted earlier.

6 Discussion and conclusion

A reliable and sufficiently accurate method has been applied to derive recent glacier area and volume changes in the Ötztal Alps out of two existing glacier inventories as well as high-resolution LIDAR-DEMs. An important prerequisite is the application of consistent methods and decision rules (e.g. concerning ice divides, debris-covered areas, snow and ambiguous areas), in order to create comparable inventories.

Absolute as well as relative glacier area losses have increased stronger at low elevations than at high elevation (Fig. 5b). This may be due to changes in energy balance (e.g. air temperature, albedo) and in the fraction of solid precipitation of the overall precipitation that have a stronger effect on glacier changes at low elevations since these regions are generally closer to melting conditions. Furthermore, ice motion slowed down significantly during recent years, which subsequently reduces dynamic ice supply to glacier tongues (e.g. Abermann et al., 2007).

Additional information such as annual length change and mass balance measurements has been used to derive estimates for acceleration factors of net area and volume loss. They have been compared with factors that would result if no temporal reduction is performed. Both methods follow a similar trend over the size classes with much higher values when considering the whole period (including the advance). The relative distribution over all size classes is not significantly affected by the temporal reduction but the absolute values of the factors. Their absolute value though has a different meaning: we interpret the overall acceleration factors as describing altered glacier behaviour in general, whereas the temporally reduced factors are intended to reveal information on an accelerated glacier retreat.

The limited number of glaciers for which additional information (length change and mass balance measurements) exists and thus the unequal distribution over the size classes of continuously measured glaciers allows only generalised estimates for the individual periods, but no year-by-year analysis of glacier retreat. Therefore no annual area and volume changes for a selected year can be calculated for the entire inventory: rather longer-term trends (e.g. decadal means of annual changes) were compared.

Length and area changes are linked but do not necessarily follow exactly the same pattern. It depends on the topography of the individual glacier how length and area respond to a climatic change but the sign of this evolution is likely to be the same. Since we only extract the temporal evolution of the length change and relate them qualitatively to the observed area change we feel confident enough to do so.

The fact that for the smallest class ($<0.1\,\text{km}^2$) retreat has decelerated in terms of area change suggests that these glaciers are now in a state closer to equilibrium than in the previous period. Differences in area change between the two periods are very large in this class (Table 1, Figs. 7 and 8) therefore we draw this conclusion even though interpretation uncertainties are larger for small glaciers.

Comparison of acceleration factors reveals that those connected with volume changes (F_{V_retr}, $F_{\bar{z}\_\text{retr}}$) are significantly larger than those connected with area changes (F_{A_retr}, $F_{A\%\_\text{retr}}$). This means that mean annual volume and mean thickness changes have increased more than mean annual area changes across all size classes compared to the respective values for the period 1969–1997. Taking an extensive data set of ice thickness measurements in the study area into consideration we found that the larger valley glaciers still have a considerable ice thickness at low elevations (a typical value for glaciers within the study area between 1 and 5 km^2 is 60 m at the tongue around 2700 m a.s.l., and up to 150 m for larger glaciers; Span et al., 2005; Fischer et al., 2007). This geometric arrangement results in the fact that mean annual volume or mean thickness changes (thick glacier tongues melt down) increase strongly while mean annual area changes do not increase significantly. This particularly affects the size class of glaciers between 0.5 and 1 km^2 where the mean relative annual area change did not change ($F_{A\%\_\text{retr}}=0.96$) but the mean annual volume change as well as the mean annual thickness change more than doubled ($F_{V\_\text{retr}}=2.16$, $F_{\bar{z}\_\text{retr}}=2.53$). Also the two smallest classes show larger differences between the factors connected with area and the ones connected with volume changes. It can be anticipated that this trend may reverse as soon as the tongues become thin enough to melt away and lose large areas within a few years without losing a lot of volume. This point in time is individual for each glacier and should be evaluated further together with ice volume data and an energy balance model.

A future application of this dataset could be to compare geodetically derived cumulative mass balances with glaciologically measured ones and assess the possible causes for these differences. Furthermore, this dataset is planned to be extended for other glacier-covered areas in the eastern Alps and will thus allow to extract regional differences of glacier reaction to a changing climate.

Acknowledgements. This study was funded by the Commission for Geophysical Research, Austrian Academy of Science. The LIDAR DEM 2006 was acquired by the Regional Government of Tyrol. The authors would like to thank the Commisssion for Glaciology at the Bavarian Academy of Sciences and Humanities for providing the mass balance data of Vernagtferner, C. Knoll for his comments and L. Raso and E. Dryland for proof-reading the paper as well as M. Attwenger for providing information on the DEM. L. Braun, I. Evans, M. Pelto and one anonymous referee are gratefully acknowledged for constructive remarks and useful suggestions, which improved the manuscript considerably.

Edited by: G. H. Gudmundsson

References

Abermann, J., Lambrecht, A., and Schneider, H.: Analysis of surface elevation changes on Kesselwand glacier – Comparison of different methods, Zeitschrift für Gletscherkunde und Glazialgeologie, 41, 147–168, 2007.

Abermann, J., Fischer, A., Lambrecht, A., and Geist, T.: Glacier mapping with airborne LIDAR and multi-temporal DEMs, The Cryosphere, submitted, 2009.

Andreassen, L. M., Paul, F., Kääb, A., and Hausberg, J. E.: Landsat-derived glacier inventory for Jotunheimen, Norway, and deduced glacier changes since the 1930s, The Cryosphere, 2, 131–145, 2008,
http://www.the-cryosphere-discuss.net/2/131/2008/.

Baltsavias, E. P., Favey, E., Bauder, A., Bösch, H., and Pateraki, M.: Digital Surface Modelling by Airborne Laser Scanning and

Digital Photogrammetry for Glacier Monitoring, Photogramm. Rec., 17(98), 243–273, 2001.

Citterio, M., Diolaiuti, G., Smiraglia, C., D'Agata, C., Carnielli, T., Stella G., and Siletto, G. B.: The fluctuations of Italian glaciers during the last century: a contribution to knowledge about Alpine glacier changes, Geogr. Ann. A, 89(3), 167–184, 2007.

Dyurgerov, M. B. and Meier, M. F.: Twentieth century climate change: Evidence from small glaciers, P. Natl. Acad. Sci. USA, 97(4), 1406–1411, 2000.

Eder, K., Würländer, R., and Rentsch, H.: Digital photogrammetry for the new glacier inventory of Austria, IAPRS International Archives of Photogrammetry and Remote Sensing, 33, 1–15, 2000.

Elsberg, D. H., Harrison, W. J., Echelmeyer, K. A., and Krimmel, R. M.: Quantifying the effect of climate and surface change on glacier mass balance, J. Glaciol., 47(159), 649–658, 2001.

Endres, J.: Farborthophotokarte "Vernagtferner 1999" aus Amateur-Luftbildern, Master thesis, Technical University of Munich, Munich, 2001.

Fischer, A., Span, N., Kuhn, M., Butschek, M.: Radarmessungen der Eisdicke österreichischer Gletscher. Band II: Messungen 1999 bis 2006, Zentralanstalt für Meteorologie und Geodynamik, Wien, Österreichische Beiträge zu Meteorologie und Geophysik, 39, 142 pp., 2007.

Fischer, A. and Markl G.: Mass balance measurements on Hintereisferner, Kesselwandferner and Jamtalferner 2003 to 2006: database and results, Zeitschrift für Gletscherkunde und Glazialgeologie, 42(1), 47–83, 2009.

Fischer, A.: Glaciers and climate change: Interpretation of 50 years of direct mass balance of Hintereisferner, Austria, Global Planet. Change, submitted, 2009.

Fliri, F.: Das Klima im Raume von Tirol, Universitätsverlag Wagner, Innsbruck-München, 1975.

Geist, T., Elvehoy, H., Jackson, M., and Stötter, J.: Investigations on intra-annual elevation changes using multitemporal airborne laser scanning data – case study Engabreen, Norway, Ann. Glaciol., 42, 195–201, 2005.

Geist, T. and Stötter, J.: Documentation of glacier surface elevation change with multi-temporal airborne laser scanner data - case study: Hintereisferner and Kesselwandferner, Tyrol, Austria, Zeitschrift für Gletscherkunde und Glazialgeologie, 41, 77–106, 2007.

Gross, G.: Der Flächenverlust der Gletscher in Österreich 1850-1920-1969, Zeitschrift für Gletscherkunde und Glazialgeologie, 23(2), 131–141, 1987.

Haeberli, W., Frauenfelder, R., Hoelzle, M., and Maisch, M.: On rates and acceleration trends of global glacier mass changes, Geogr. Ann. A, 81(4), 585–591, 1999.

Heipke, C., Rentsch, H., Rentsch, M., and Würländer, R.: The digital orthophoto map Vernagtferner 1990, Zeitschrift für Gletscherkunde und Glazialgeologie, 30, 109–117, 1994.

Kääb, A., Paul, F., Maisch, M., and Häberli, W.: The new remote-sensing-derived Swiss Glacier Inventory: II. First results, Ann. Glaciol., 34, 362–366, 2002.

Kennett, M. and Eiken, T.: Airborne measurement of glacier surface elevation by scanning laser altimeter, Ann. Glaciol., 24, 293–296, 1997.

Knoll, C. and Kerschner, H.: A glacier inventory for South Tyrol, Italy, based on airborne laser scanner data, Ann. Glaciol., 50(53), 46–52, 2009.

Kodde, M., Pfeiffer, N., Gorte, B., Geist, T., and Höfle, B.: Automatic Glacier Surface Analysis from Airborne Laser Scanning, ISPRS Workshop Laser Scanning 2007, XXXVI, Part 3/W52, 221–226, 2007.

Kuhn, M.: Begleitworte zur Karte des Hintereisferners 1979, 1:10 000, Zeitschrift für Gletscherkunde und Glazialgeologie, 16(1), 117–124, 1979.

Kuhn, M., Markl, G., Kaser, G., Nickus, U., Obleitner, F., and Schneider, H.: Fluctuations of climate and mass balance: Different responses of two adjacent glaciers, Zeitschrift für Gletscherkunde und Glazialgeologie, 21, 409–461, 1985.

Kuhn, M., Dreiseitl, E., Hofinger, S., Markl, G., Span, N., and Kaser, G.: Measurements and models of the mass balance of Hintereisferner, Geogr. Ann. A, 81(4), 659–670, 1999.

Kuhn, M., Lambrecht, A., Abermann, J., Patzelt, G., and Gross, G.: Projektbericht 10. Die österreichischen Gletscher 1998 und 1969, Flächen- und Volumenänderungen, Verlag der österreichischen Akademie der Wissenschaften, Wien, 128 pp., 2009.

Lambrecht, A. and Kuhn, M.: Glacier changes in the Austrian Alps during the last three decades, derived from the new Austrian glacier inventory, Ann. Glaciol., 46, 177–184, 2007.

Lemke, P., Ren, J., Alley, R. J., et al.: Observations: Changes in Snow, Ice and Frozen Ground, in: Climate Change 2007: The Physical Science Basis. Contribution of Working Group I to the Fourth Assessment Report of the Intergovernmental Panel on Climate Change, edited by: Solomon, S., Qin, D., Manning, M. et al., Cambridge University Press, Cambridge, United Kingdom and New York, NY, USA, 2007.

Oerlemans, J.: Extracting a climate signal from 169 glacier records, Science, 308, 675–677, 2005.

Patzelt, G.: The Austrian glacier inventory: status and first results, Riederalp Workshop 1978 – World Glacier Inventory, IAHS, 1980.

Patzelt, G.: The period of glacier advances in the Alps, 1960 to 1985, Zeitschrift für Gletscherkunde und Glazialgeologie, 21, 403–407, 1985.

Patzelt, G.: Gletscherbericht 2003/2004. Sammelbericht über die Gletschermessungen des Oesterreichischen Alpenvereins im Jahre 2004, Mitteilungen des Oesterreichischen Alpenvereins, 60(130), 24–31, 2005.

Patzelt, G.: Gletscherbericht 2004/2005. Sammelbericht über die Gletschermessungen des Österreichischen Alpenvereins im Jahre 2005, Bergauf, 2, 6–11, 2006.

Paul, F., Kääb, A., Maisch, M., Kellenberger, T. W., and Häberli, W.: The new remote-sensing-derived Swiss Glacier Inventory: I. methods, Ann. Glaciol., 34, 355–361, 2002a.

Paul, F.: Changes in glacier area in Tyrol, Austria, between 1969 and 1992 derived from Landsat 5 Thematic Mapper and Austrian Glacier inventory data, Int. J. Remote Sens., 23(4), 787–799, 2002b.

Paul, F., Kääb, A., Maisch, M., Kellenberger, T., and Häberli, W.: Rapid disintegration of Alpine glaciers observed with satellite data, Geophys. Res. Lett., 31, L21402, doi:10.1029/2004GL020816, 2004.

Paul, F. and Haeberli, W.: Spatial variability of glacier elevation changes in the Swiss Alps obtained from two digital elevation models, Geophys. Res. Lett., 35, L21502, doi:10.1029/2008GL034718, 2008.

Paul, F., Barry, R., Cogley, G., Frey, H., Haeberli, W., Ohmura, A., Ommanney, S., Raup, B., Rivera, A., and Zemp, M.: Recommendations for the compilation of glacier inventory data from digital sources, Ann. Glaciol., submitted, 2009.

Pelto, M. S.: The current disequilibrium of North Cascade Glaciers, Hydrol. Process., 20, 769–779, 2006.

Schneider, C., Schnirch, M., Acuña, C., Casassa, G., and Kilian, R.: Glacier inventory of the Gran Campo Nevado Ice Cap in the Southern Andes and glacier changes observed during recent decades, Global Planet. Change, 59(1–4), 87–100, 2007.

Span, N., Fischer, A., Kuhn, M., Massimo, M., and Butschek, M.: Radarmessungen der Eisdicke österreichischer Gletscher; Band I: Messungen 1995 bis 1998, Zentralanstalt für Meteorologie und Geodynamik, Wien, Österreichische Beiträge zu Meteorologie und Geophysik, 33, 145 pp., 2005.

UNESCO: Perennial ice and snow masses: a guide for compilation and assemblage of data for a world inventory, UNESCO/IASH, Technical Paper Hydrology 1, 1970.

Würländer, R. and Eder, K.: Leistungsfähigkeit aktueller photogrammetrischer Auswertemethoden zum Aufbau eines digitalen Gletscherkatasters, Zeitschrift für Gletscherkunde und Glazialgeologie, 35, 167–185, 1998.

A4: POSTER I: TOWARDS A THIRD AUSTRIAN GLACIER INVENTORY: FIRST RESULTS AND A CLIMATIC INTERPRETATION

By: Abermann, J., B. Seiser and A. Fischer. *Presented at the European Geoscience's General Assembly 2010*, Vienna, Austria.

Towards a third Austrian glacier inventory: First results and a climatic interpretation

IMGI

J. Abermann[1,2]*, B. Seiser[2], A. Fischer[2]

*jakob.abermann@uibk.ac.at

[1] Austrian Academy of Sciences, Comm. for Geophysical Research
[2] Institute of Meteorology and Geophysics, University of Innsbruck

ÖAW Austrian Academy of Sciences

1 MOTIVATION

➢strong glacier area and volume changes observed in the Alps and in large parts of the world in the past decades
➢Two complete glacier inventories exist in Austria
➢2006: LIDAR-Data – Database for a new inventory

RESEARCH TASKS:

✓ ***Quantification of the accelerated glacier retreat***
✓ ***Investigation of two adjacent mountain groups***
- ***Similar but not same climate***
- ***Different glacier characteristics/ topography***

✓ ***Extract regional differences of glacier mass changes and link them to climate data?***

2 STUDY AREA

➢Austrian Alps, central Europe (Fig. 1)
➢Glacier distribution: 470 km² (1998), approx. 900 glaciers (2000 to 3800m a.s.l.)
➢Ötztal and Stubai Alps : 2006: 165 km² (~40% of Austrian glacier cover)

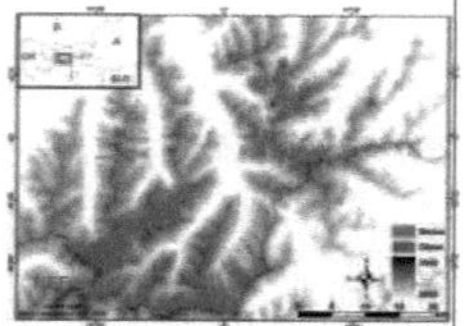

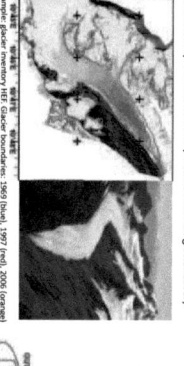

Fig. 1: Study area: Ötztal glaciers (blue) and Stubai glaciers (red). The location of Hintereisferner (HEF, Fig. 2) is marked with a red arrow.

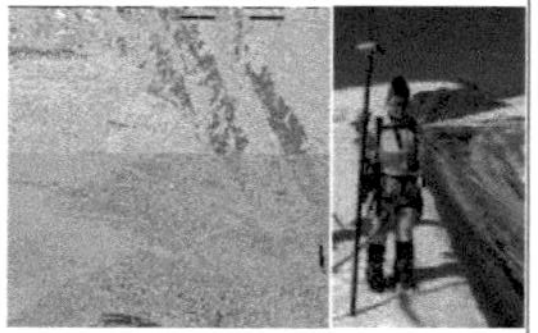

Fig. 2: Data example: glacier inventory HEF. Glacier boundaries: 1969 (blue), 1997 (red), 2006 (orange) as well as thickness changes 1997 – 2006 (left). Aerial photograph from September 2008 (right).

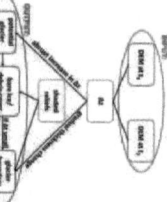

Fig. 3: Workflow of the glacier mapping procedure with multitemporal (high-resolution) DEMs.

3 DATA AND METHODS

I)Glacier inventories:
➢2 complete glacier inventories (1969[1] and 1998[2], e.g.Fig.2)
➢2006 LIDAR-survey
→cell-size: 1m
→Accuracies: +/-0.3 horizontal; +/-0.1 vertical
→Glacier boundary delineation including multi-temporal DEM information (Fig. 3 and 4)[3]
→DEMs
→ΔV

II)Climate data
➢ERA40 reanalysis data, 6-hourly [4]
➢gridded precipitation dataset, monthly [5]

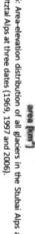

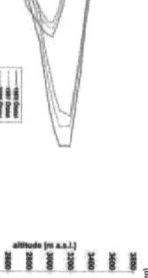

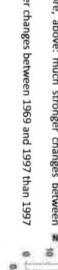

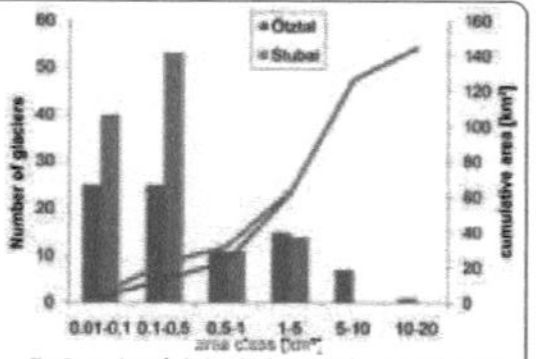

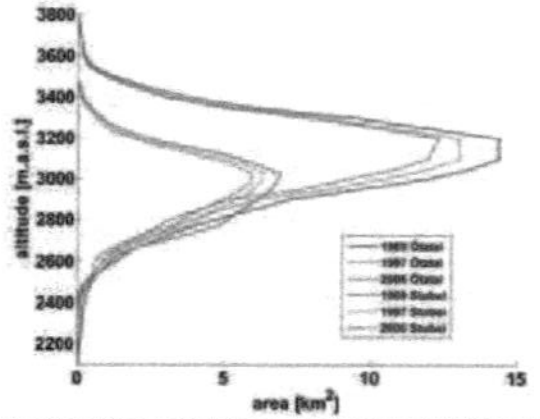

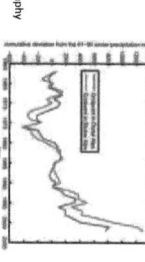

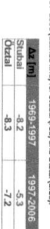

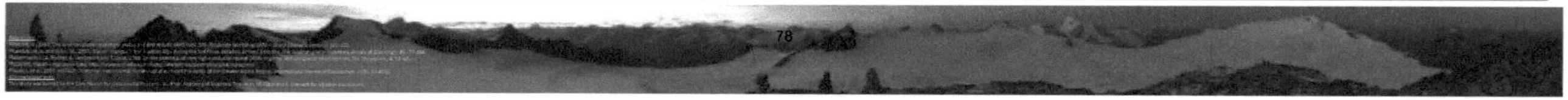

Fig. 4: The glacier margin of Vernagtferner with the LIDAR-derived boundary (solid) and the geodetically measured one (crosses). The GPS-survey of the glacier boundary for ground-truthing.

4 RESULTS

➢Stubai: mainly small glaciers
➢Ötztal: glaciers >5km² make more than 50% of total area (Fig.5)
➢Median elevation of glacier cover about 250 m higher in Ötztal than Stubai (Fig. 6)
➢Maximal absolute area changes coincide with maximal glacier cover
➢Ötztal: below 2800 m same absolute glacier area between 1997 and 2006 as in 28 years before, above: much stronger changes between 1969 and 1997
➢Stubai: generally stronger changes between 1969 and 1997 than 1997 – 2006 (Fig.7)

Fig. 5: number of glaciers in the respective mountain group (bars, left axis). Cumulative area (lines, right axis).

Fig. 6: Area-elevation distribution of all glaciers in the Stubai Alps and the Ötztal Alps at three dates (1969, 1997 and 2006).

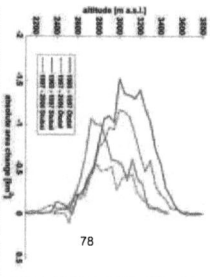

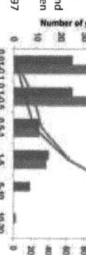

Fig. 7: Absolute area changes between 1969 and 1997 (solid line) and 1997-2006 (dashed line) for Stubai (red) and Ötztal (blue).

➢Differences reflected in mean thickness change (Tab. 1):
→First period: similar changes
→Second period: stronger negative changes in Ötztal than Stubai

Δz [m]	1969-1997	1997-2006
Stubai	-8.2	-5.3
Ötztal	-8.3	-7.2

Tab. 1: Mean thickness changes between 1969 and 1997 and 1997-2006 for Stubai (red) and Ötztal (blue).

5 CLIMATIC CONSIDERATIONS

➢South-North gradient of glacier changes explainable??
→Temperature evolution similar (ERA40, weather stations)
→cumulative deviations from mean precipitation show (Fig. 8):
- Similar anomalies until ~1990
- Stronger positive anomalies of gridpoint Stubai since then

Fig. 8: Cumulative anomalies of mean winter precipitation for a gridpoint in Stubai (red) and Ötztal (blue).

6 CONCLUSIONS AND OUTLOOK

✓Glacier inventory studies with LIDAR-DEMs
✓Regional differences of glacier changes significant – different topography
✓Further spatial analysis (other regions, larger scale) planned
✓Investigations of changed synoptic patterns

78

A5: POSTER II: SYNCHRONEOUS GLACIER RETREAT NORTH AND SOUTH OF A CENTRAL-ALPINE MOUNTAIN DIVIDE

By: Abermann, J., C. Knoll, H. Kerschner and A. Lambrecht. *Presented at the Alpine Glaciology Meeting 2009*, Innsbruck, Austria.

Synchroneous glacier retreat North and South of a Central-Alpine Mountain divide?

J. Abermann[1,2]*, C. Knoll[3], H. Kerschner[3], A. Lambrecht[2]
*jakob.abermann@uibk.ac.at

[1] Austrian Academy of Sciences, Comm. f. Geoph. Research, Vienna
[2] Institute of Meteorology and Geophysics, University of Innsbruck
[3] Institute of Geography, University of Innsbruck

INTRODUCTION
We compare the evolution of glacier area and volume changes along the alpine main ridge in the Ötztal Alps and Texel group using glacier inventory data from 1997 and 2006 and the reconstructed Little Ice Age (LIA) extent (approx. 1850).

STUDY AREA
The southern Ötztal Alps and the Texel Group are situated around 47°N and 11°E (Fig. 1) along the border of Austria and Italy. Annual precipitation is rather low with typical values of about 700 mm/a in 1900 m a.s.l. Glaciers on the northern slope reach down to 2100 m and on the southern slope to 2400 m.

DATA
The 1997 glacier inventories were compiled with digital photogrammetry, while Airborne Laser Scanning (ALS) were used for the inventory for 2006 (Fig. 2). The LIA extent was drawn with the help of moraines, trimlines and historical maps (see Knoll and Kerschner (2009), Gross (1987)).

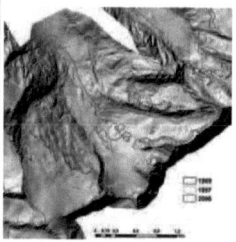

Figure 2: Rotmoosferner and its surrounding glaciers with the hillshade of the DEM and glacier boundaries for 1969, 1997 and 2006.

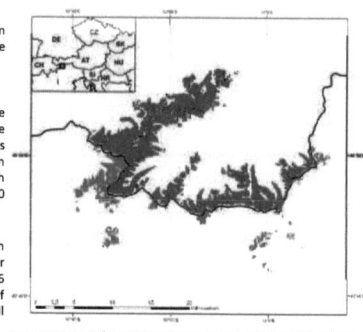

Figure 1: The study area Ötztal Alps and Texel Group. Glaciers north (blue) and south (red) of the main alpine divide.

Figure 3: Kreuzferner and Hochjochferner, September 2008.

	Relative area change [%]		mean thickness change [m]	
	N	S	N	S
1850 - 1997	-37.3	-61.8	-48.1	-30.2
1997 - 2006	-8.3	-12.9	-8.2	-9.0

Table 1: Relative glacier area and mean thickness changes for 1850 – 1997 and 1997 – 2006 for the northern (N) and southern (S) part of the study area.

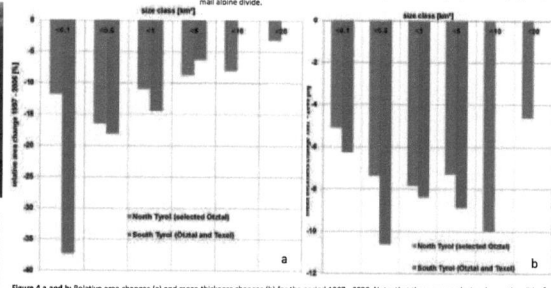

Figure 4 a and b: Relative area changes (a) and mean thickness changes (b) for the period 1997 - 2006. Note, that there are no glaciers larger than 5 km² in South Tyrol.

RESULTS
Area loss on the southern side was more pronounced than on the northern side. Between the LIA and 1997, mean thickness change (volume change divided by a mean area of the period) was larger in the north than in the south, while it was larger in the south between 1997 and 2006 (Tab. 1 and Fig. 4a). For the 1997 to 2006 period, South-Tyrolean glaciers showed a larger relative area change except for the class of 1-5 km². Very small glaciers show the largest differences. Mean thickness change is larger in the south than in the north for 1997 – 2006 (Fig. 4b).

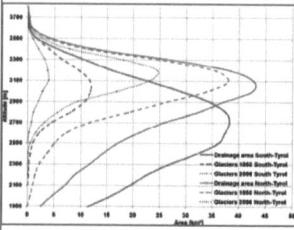

Figure 5: Area – Altitude distribution of the drainage area as a whole (solid), the glacier covered area for the LIA-maximum (dashed) and 2006 (dotted).

Figure 6: Matscherferner with Weißkugel. Gepatschferner's large ice-plateau is visible in the background

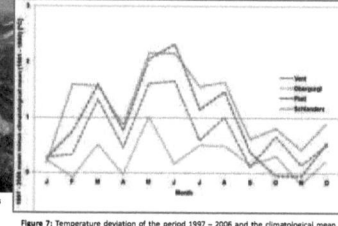

Figure 7: Temperature deviation of the period 1997 – 2006 and the climatological mean (1961 - 1990) [°C] for selected weather stations north (blue) and south (red) of the main divide.

DISCUSSION AND CONCLUSIONS
The pronounced relative area loss in the south relative to the north since the LIA is a consequence of the area-altitude distribution of the mountain ranges and hence the glacier covered areas (Fig. 5). The northern parts show larger areas at higher elevations than the south. Due to glacier dynamics, this leads to larger and thicker ice masses at lower altitudes during the LIA. This explains the stronger reduction in ice thickness between LIA and 1997 in the north. Additionally, volume change is limited through the initial ice thickness. The larger and thicker glaciers in the north allow a stronger thickness loss.
For the 1997 - 2006 period, stronger area and thickness changes in the south may be due to climatological factors (Fig. 7). Precipitation data from the HISTALP database (Auer et al., 2005) show generally slightly higher winter precipitation in the north but no significant changes in the precipitation pattern (not shown). The summer temperature rise of the last decade was clearly higher in the south at selected weather stations (Fig. 7).

Literature
Abermann, J., Lambrecht, A., Fischer, A. and Kuhn, M., 2009: Quantifying changes and trends in glacier area and volume of a study area in the Ötztal Alps. In prep.
Auer, I. et al., 2007. HISTALP - historical instrumental climatological surface time series of the Greater Alpine Region, International Journal of Climatology, 27(1), 17-4166.
Gross, G. 1987: Der Flächenverlust der Gletscher in Österreich 1850–1920–1969. Zeitschrift für Gletscherkunde und Glazialgeologie 23(2), 131-141.
Knoll, C. and Kerschner, H., 2009 (submitted). A glacier inventory for South Tyrol, Italy, based on airborne laser scanner data. Annals of Glaciology, 53.
Lambrecht, A. and Kuhn, M. 2007. Glacier changes in the Austrian Alps during the last three decades, derived from the new Austrian glacier inventory. Annals of Glaciology, 46: 177-184.

Acknowledgements
This work was funded by the Commission of Geophysical Research, Academy of Sciences, Vienna and the Vizerectorate for Research of the University of Innsbruck

A6: PAPER IV: A RECONSTRUCTION OF ANNUAL MASS BALANCES OF AUSTRIA'S GLACIERS FROM 1969 TO 1998

By: Abermann, J., M. Kuhn and A. Fischer. *Published in Annals of Glaciology*, *59*(51), 127-134.

A reconstruction of annual mass balances of Austria's glaciers from 1969 to 1998

J. ABERMANN,[1,2] M. KUHN,[1,2] A. FISCHER[2]

[1]*Commission for Geophysical Research, Austrian Academy of Sciences, Dr-Ignaz-Seipel-Platz 2, A-1010 Vienna, Austria
E-mail: jakob.abermann@uibk.ac.at*
[2]*Institute of Meteorology and Geophysics, University of Innsbruck, Innrain 52, A-6020, Innsbruck, Austria*

ABSTRACT. Annual glacier mass balances are reconstructed for 96% of the Austrian glacier-covered area (451 of 470 km^2) between 1969 and 1998. The volume change derived from two complete glacier inventories (1969 and 1998) serves as the boundary condition that is aimed to be reproduced. ERA-40 reanalysis data as well as a gridded precipitation dataset (HISTALP) are used to drive a positive degree-day (PDD) model. The results are verified with four independent long-term mass-balance series. The spatial and vertical distribution of the tuning parameters is altered in order to reproduce the measured mean annual surface mass balances of selected glaciers, and a strong correlation is found between the median elevation of a glacier and the degree-day factor (DDF) at this elevation. This result implies that the lower a glacier's median elevation is, the less melt occurs at a given elevation and temperature. We attribute this to the fact that lower-altitude glaciers are generally those with more accumulation, which leads to later exposure of bare ice and a longer period of high-albedo snow cover. A further improvement of the model was achieved by making DDF a function of time as well as space. The results indicate that mean DDFs generally increase for a given date over a sequence of consecutive negative mass-balance years, which probably reflects the reduction in albedo related to that. Finally, the major drivers of the observed mass-balance evolution are investigated: summer PDD sums correlate significantly better with the observed mass-balance changes than annual PDD sums or precipitation do. This implies that annual mass balances in the study area are governed by summer temperatures.

1. INTRODUCTION

Mountain glaciers contribute considerably to observed and predicted sea-level rise (e.g. Kaser and others, 2006; Lemke and others, 2007; Meier and others, 2007). In order to estimate this contribution, it is crucial to know glacier mass balances. Mass-balance measurements are restricted to few glaciers worldwide, in total around 250 (Dyurgerov and Meier, 2005), and only about 85 data series last longer than 10 years (Braithwaite, 2002). Most glaciers for which mass balances are measured are chosen for accessibility or by coincidence. However, they might not be representative of the mass balances of all glaciers in a given catchment or mountain range, and simple extrapolation to other glaciers is known to be unreliable (Fountain and others, 2009).

Different methods have been used to model the total glacier mass balance of mountain ranges. Machguth and others (2009) used output from regional climate models (RCMs) to run an energy-balance model for all Swiss glaciers, and Schöner and Böhm (2007) applied a statistical approach to model the mass balance of a sample of Austria's glaciers as far back as the glacier maximum of the Little Ice Age. Hock and others (2009) estimate the world's glaciers' total contribution to sea-level rise. It has also been shown that reanalysis data can be used successfully to reconstruct the mass-balance history of individual glaciers (e.g. Radić and Hock, 2006; Rasmussen and Wenger, 2009).

The annual surface mass balance of 761 Austrian glaciers is reconstructed for all years between 1969 and 1998, which is an important input for run-off modelling on a catchment or larger scale. A positive degree-day (PDD) model is applied, and its performance validated. Although PDD models are generally viewed as crude because they do not resolve energy-balance terms explicitly, it has been shown in the literature that their performance is comparable to energy-balance models (Hock, 2003). The scale of the study area, and the computational simplicity required to apply the model to a large number of glaciers over a 30 year time-span, was one reason for choosing a simple approach. Furthermore, the main task is to describe measured volume changes with higher temporal resolution which can be reliably achieved with such a model. However, it is shown that it is necessary to use temporally and spatially varying degree-day factors (DDFs). In order to reproduce directly measured volume changes and their variations, one needs annual resolution, i.e. mean annual DDFs for each elevation band. While the stepwise change from low DDFs of snow and firn to the high DDFs of bare ice moves up-glacier during the season, the annual mean of DDF(h) with altitude becomes a smooth profile which is approached by a linear function of elevation in this study.

The generalizability of directly measured mass balances is addressed, and the relative importance of the main meteorological factors (temperature, precipitation) that control each year's mass balance is investigated.

2. STUDY AREA

The study area is the Austrian part of the eastern Alps (46°40′–47°35′ N, 9°50′–13°40′ E) where, in 1998, 910 glaciers covered a total area of 470 km^2 (Lambrecht and Kuhn, 2007; Kuhn and others, 2009). Figure 1 shows a map with the 1998 glacier extent and an Advanced Spaceborne Thermal Emission and Reflection Radiometer (ASTER) digital elevation model (DEM) of elevations higher than 2000 m in greyscale. The ASTER DEM is used for illustrative purposes only; our analysis was performed with photogrammetrically

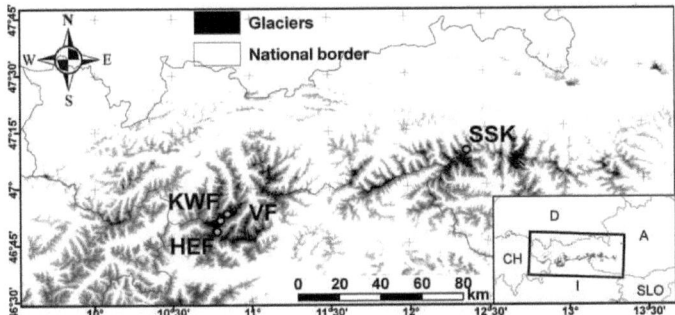

Fig. 1. Glacier cover in Austria according to the glacier inventory of 1998 (Lambrecht and Kuhn, 2007). The glaciers HEF, KWF, VF and SSK are marked. Elevations higher than 2000 m, derived from an ASTER DEM, are displayed in greyscale in the background.

derived DEMs (see section 3). Mass balances have been measured over the entire modeled time-span on four glaciers only: Hintereisferner (HEF), Kesselwandferner (KWF), Vernagtferner (VF) and Stubacher Sonnblickkees (SSK) (Schöner and Böhm, 2007). These glaciers' mass-balance series are used to validate the model; their locations are shown in Figure 1 and basic glaciological parameters are provided in Table 1. More information on the distribution of Austria's glaciers and its relationship to climate can be found in Abermann and others (2009).

3. DATA

3.1. Glaciological data

Two glacier inventories serve as glaciological input data for this study. Both were produced by aerial photogrammetry including glacier boundaries and DEMs (e.g. Gross, 1987; Lambrecht and Kuhn, 2007). Glacier area changes, which are necessary to compute annual surface mass balances, are approximated by assuming that the initial area (1969) remains constant until 1985 and then interpolating linearly to the final glacier area (1998). This pattern of area change does not account for glacier growth in the late 1970s and early 1980s but is similar to the mean glacier area changes of the study period (Abermann and others, 2009).

Volume change, ΔV, is calculated by subtracting the 1998 DEMs from the 1969 glacier extent. The total balance,

Table 1. Basic glaciological parameters for the glaciers used to validate the model: area, minimum elevation (z_{min}), maximum elevation (z_{max}), median elevation (z_{med}), main aspect of the ablation area, latitude and longitude as given in Lambrecht and Kuhn (2007) and Kuhn and others (2009) for 1998

Glacier	Area	z_{min}	z_{max}	z_{med}	Aspect	Lat. (N)	Long. (E)
	km²	m	m	m			
HEF	8.4	2400	3710	2990	NE	46°48′	10°46′
KWF	4.2	2690	3500	3140	SE	46°51′	10°48′
VF	8.8	2760	3630	3080	SE	46°53′	10°49′
SSK	1.5	2490	3030	2760	E	47°08′	12°36′

B, was calculated by

$$B = \Delta V_{1969-98}\, \rho \quad [\text{kg or } 10^{-3}\, \text{m}^3 \text{ w.e.}]. \quad (1)$$

The whole volume change is assumed to have occurred as loss of ice with a density of 900 kg m⁻³. The resulting B has thus to be taken as the upper bound of the actual mass loss, as some unknown fraction of the total volume change was lost as snow instead of ice. This uncertainty is discussed in section 6.

Directly measured mass-balance data of four glaciers spanning the entire investigation period (Table 1) were used to validate the model. Three of the glaciers are situated very close to each other in the Ötztal valley: HEF, KWF, VF (Hoinkes and Lang, 1962; Kuhn and others, 1985, 1999; Fischer and Markl, 2008; Fischer, 2010) and VF (Reinwarth and Escher-Vetter, 1999; Escher-Vetter and others, 2009). SSK (Slupetzky, 1989, 2003; WGMS, 2007) is located in the Hohe Tauern and thus in a somewhat different climate (generally more precipitation; Abermann and others, 2011).

3.2. Meteorological data

The model uses temperature, T, equivalent-potential temperature, θ_e, and geopotential, Φ, and precipitation as meteorological inputs. T, θ_e and Φ and precipitation are taken from the European Centre for Medium-Range Weather Forecasts (ECMWF) ERA-40 reanalysis project (Uppala and others, 2005). T, θ_e and Φ are taken from the 600, 700 and 850 hPa level; precipitation data are the 6 hour forecast from the total precipitation field at the surface. The ERA-40 reanalysis dataset is a dynamically consistent, three-dimensional, gridded dataset with a resolution of 1.25° that combines meteorological and satellite observations with a numerical weather-forecast model and covers the period 1957–2002.

The monthly Alps-wide precipitation dataset of the HISTALP project has a spatial resolution of 10′ of arc and includes homogenized weather-station data (Efthymiadis and others, 2006). The dataset was generated with data from valley stations only (Auer and others, 2005) due to well-known problems of precipitation measurement at high altitudes (e.g. Frei and Schär, 1998). It therefore does not account for an increase in precipitation with altitude and underestimates high-altitude precipitation.

Generally, precipitation anomalies have the same sign at high- and at low-altitude stations (Auer and others, 2005; Böhm and others, 2008). In order to account for the increase in precipitation with elevation, $12\,mm\,(100\,m)^{-1}$ were added to each month's precipitation sum up to a maximum elevation of 3300 m, above which this value is assumed to remain constant. This vertical gradient is larger than annual gradients given by Kuhn and others (1982) and Kuhn (2003). Winter precipitation mainly consists of advective precipitation (unlike summer precipitation, where convection plays an important role); larger vertical gradients for winter precipitation are therefore expected for winter precipitation than for annual precipitation. Since accumulation is determined by winter precipitation, the chosen gradient best represented the observed vertical balance profile of selected glaciers.

HISTALP and ERA-40 precipitation are combined with the ERA-40 θ_e to incorporate a spatially (HISTALP) and temporally (ERA-40) highly resolved precipitation dataset into the model. The higher-spatial-resolution HISTALP dataset is used to determine the total value for a month at a gridpoint and then split between the 6 hourly time-steps given by the ERA-40 precipitation data and weighted accordingly. As an example, let HISTALP's total November precipitation at a glacier's gridpoint be 90 mm and let ERA-40 show that 30% of the month's precipitation fell on 6 November and 70% on 20 November. The monthly precipitation is then divided into two time-steps: 27 mm on 6 November and 63 mm on 20 November.

The rain–snow boundary is estimated by applying the regression function of Steinacker (1983) who related the rain–snow boundary empirically to the equivalent-potential temperature at 850 hPa. He found a relationship of the rain–snow boundary $z_{r/s}$ of

$$z_{r/s} = \left(\frac{\theta_e - 12}{1.2}\right) \times 100, \quad (2)$$

where the equivalent-potential temperature at 850 hPa is inserted in °C (Steinacker, 1983). If, at a certain point in time, θ_e at 850 hPa is 42°C the rain–snow boundary is assumed to be at 2500 m, applying Equation (2).

4. THE MODEL

A PDD model is applied as widely used in hydrological modelling (e.g. Hock, 2003, 2005; Kuhn, 2003; Huss and others, 2008 and references therein). To estimate the vertical temperature distribution at a certain point in time, temperatures at 600, 700 and 850 hPa and the geopotential there are used. All glacier-covered areas in the study area lie between these pressure levels at all points in time. The geopotential of a certain pressure level, Φ, is converted into geopotential height, z_{gp}, by applying

$$z_{gp} \approx \frac{\Phi}{\bar{g}} \quad [m], \quad (3)$$

where $\bar{g}$ is the mean gravity acceleration (9.81 m s^{-2}). z_{gp} is assumed to be equal to the geometrical height, z, with a satisfactory approximation (Stull, 2000).

If z_1 is located between the 700 and 850 hPa levels, for instance, the temperature there, T_{z1}, is

$$T_{z1} = \frac{T_{700} - T_{850}}{z_{700} - z_{850}}(z_1 - z_{700}) + T_{700} \quad [K]. \quad (4)$$

The same applies analogously for z between 600 and 700 hPa.

To calculate ablation, the relation

$$b_{abl}(x, y, z, t) = DDF(x, y, z, t) T_{>0, day}(x, y, z, t) \quad [mm\,w.e.] \quad (5)$$

is used, where DDF is used for calibration and $T_{>0,day}$ is the daily mean temperature above 0°C.

Accumulation is estimated by

$$b_{acc}(x, y, z, t) = P(x, y, z, t) R_{0/1}(x, y, z, t) \quad [mm\,w.e.], \quad (6)$$

where P is precipitation and $R_{0/1}$ is a value that indicates whether the precipitation has fallen as snow or rain and that can only be 0 (rain) or 1 (snow). If $z < z_{r/s}$, it is 0, otherwise it is 1. Liquid precipitation is assumed not to contribute to glacier mass gain. Both ablation and accumulation are computed with a 6 hour time-step. Their sum gives the specific mass balance, b.

The spatial and vertical dependence of DDF is a crucial point in the model and needs explanation. In the following, the vertical dependence of DDF (dDDF/dz) is introduced first, and then the spatial dependence.

An energy-balance model, SOMARS (Simulation Of glacier surface Mass balance And Related Sub-surface processes; Greuell and Konzelmann, 1994), was applied and validated with in situ measurements of the snow water equivalent at two locations on HEF by Schrott (2006). With the results of this model (specific balance) an ablation season's mean DDF at these two weather stations (2650 and 3050 m) is calculated. DDF varies vertically by –0.32 mm w.e. (Kd)$^{-1}$ per 100 m of elevation. Over a 1000 m height interval this amounts to a difference of –3.2 mm (Kd)$^{-1}$; for example, the lower end may have a DDF of 7.2 mm (Kd)$^{-1}$ (low albedo) and the upper end one of 4 mm (Kd)$^{-1}$ (high albedo). This vertical gradient of DDFs is extrapolated and used as the vertical dependence of DDF in the model. The vertically changing DDFs are in line with theoretical considerations by Hock (2003). On average over a melt season, the albedo of lower areas is smaller, so, for a given temperature, the energy added to a surface is higher, as more shortwave radiation is absorbed. This is independent of temperature and would therefore not be reflected in a temperature-index approach with a constant DDF for the whole glacier.

In addition to the vertically varying DDFs, a spatial dependence is introduced. This is found by running the model iteratively with dDDF/dz as described above and with the condition that we must obtain the measured B of each glacier as it had been previously calculated from the glacier inventories. Each individual glacier thus has its own set of DDFs that best represents the measured B. DDFs vary considerably among the individual glaciers and are discussed below.

5. RESULTS

The results of the first calibrated model run (run I) for HEF are shown in Figure 2a. It should be noted that the area changes for the modelled balances are approximated as explained above, whereas the area changes for the measured balances are adjusted each year. This approximation is made so that the method may be applied to unmeasured glaciers as well. Statistical values in Table 2 indicate that a similar quality of results is achieved for the three other glaciers where cross-

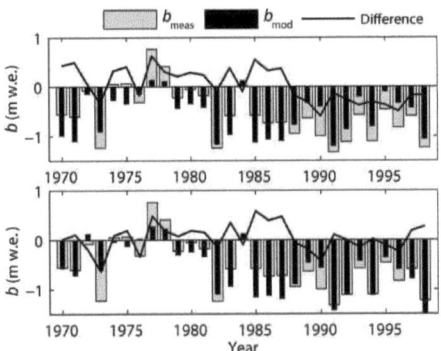

Fig. 2. Measured (b_{meas}) and modelled mean specific surface mass balance (b_{mod}) at HEF for two different model runs. (a) is calculated with constant DDFs over time; in (b), a temporally rising DDF value is introduced based on findings of Figure 3.

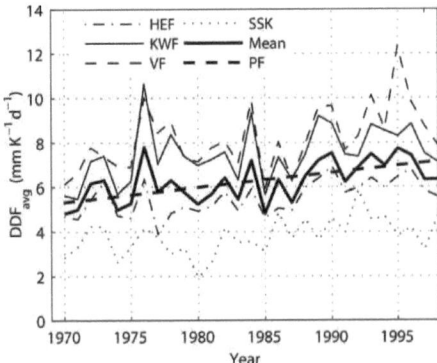

Fig. 3. Mean area- and elevation-weighted DDF that best reproduces the measured annual surface mass balance of HEF, KWF, VF and SSK and these glaciers' mean as calculated according to Equation (4). PF is the polynomial second-order fit that is calculated in order to represent the mean of the glacier's individual values.

validation data (i.e. measured mass balance) exist. Figure 2a shows that the model approximates the pattern of glacier change. There is a clear temporal dependence of the differences between measured and modelled balance. Until about 1987, measured balances are more positive than modelled balances; after 1987, this trend is reversed.

This result is investigated further by asking: which annual values do DDFs have to have averaged over an ablation season to reproduce the measured mass balance of each year on the four glaciers with continuous mass-balance data? This is shown in Figure 3 where mean annual DDFs are plotted over time. The average DDFs are calculated as

$$\text{DDF}_{avg} = \frac{B_{ann,meas} - B_{acc,mod}}{\sum \text{PDD}} \quad [m(Kd)^{-1}]. \quad (7)$$

Figure 3 shows that there is an increase in DDFs over the observation period for all glaciers for which measured mass-balance data exist: at a given temperature, more melt occurs in later than in earlier years. The increasing trend is statistically significant and its magnitude is similar for the four glaciers with complete mass-balance data. The absolute values of mean DDFs are largest for VF, followed closely by

Table 2. Mean absolute difference, MD_{abs}, and the mean signed difference, MD, between annually measured and modelled balances, the correlation coefficient, R, and the standard deviation, STD, of the differences for two model runs (run I: constant DDFs; run II: temporally changing DDFs following the findings of Fig. 3) for 1969–98. MD_{abs}, MD and STD are in m w.e., R is dimensionless

Glacier	Run I				Run II			
	MD_{abs}	MD	R	STD	MD_{abs}	MD	R	STD
HEF	0.31	−0.03	0.75	0.35	0.22	−0.05	0.85	0.28
KWF	0.33	−0.16	0.70	0.37	0.30	−0.17	0.80	0.35
VF	0.34	0.10	0.60	0.44	0.28	0.08	0.75	0.38
SSK	0.41	0.10	0.72	0.54	0.35	0.07	0.81	0.45

KWF, significantly lower for HEF and lowest for SSK. For the three glaciers of the Ötztal Alps (HEF, KWF, VF), these differences are likely due to the influence of shortwave radiation. The south-facing VF and southeast-facing KWF receive more shortwave radiation than northeast-facing HEF (Table 1) and therefore have higher DDFs, a result that is consistent with previous research (e.g. Hock, 2005). SSK also provides a valuable comparison: it is located in a different climate with significantly more precipitation (Abermann and others, 2011). More precipitation may cause a higher mean albedo and thus lower mean DDFs. We calculated a second-order polynomial fit (PF) that best represents the mean of all DDF increases, and found that the overall increase is ~1.5 mm (Kd)$^{-1}$ over the 29 years observed.

In the second model run (run II), we introduce a time dependence of DDF for each glacier. The best-fitting set of DDFs as found in run I is altered with time by the rate determined by the polynomial increase according to Figure 3. This altered DDF (DDF_{II}) at any point in time of the study period ($date_s$), which has to be inserted as a serial date number with 0 at 1 January 0000, increasing by 1 each day (amounting for example to 719529 for 1 January 1970), can be expressed by the best-fitting DDF as determined in run I (DDF_I):

$$DDF_{II}(date_s) = DDF_I - (3.306 \times 10^{-13})date_s^2 + (6.439 \times 10^{-7})date_s - 0.2930. \quad (8)$$

DDF is now a four-dimensional (3-D: run I; 4-D: run II) function of space and time. Figure 2b shows modelled balances with temporally varying DDFs and measured balances for HEF, as well as their differences. Both the trend and the annual values are better modelled than in the previous run. This is also shown in Table 2 in the mean absolute differences (MD_{abs}), mean signed differences (MD), standard deviations (STD) of the differences between measured and modelled balances and the correlation coefficients, R, between measured and modelled balances. Without exception, the second run gave better results:

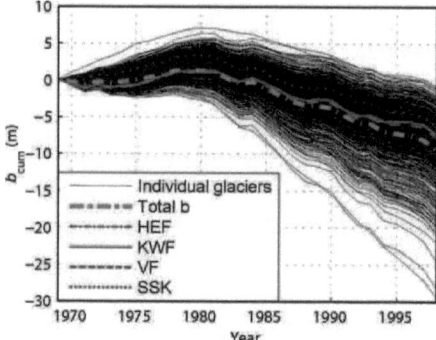

Fig. 4. Cumulative mean specific surface mass balance of all modelled glaciers in Austria (black lines) individually; the green line shows their total cumulative mean specific mass balance. The red lines show modelled balances of the glaciers where direct mass-balance measurements exist.

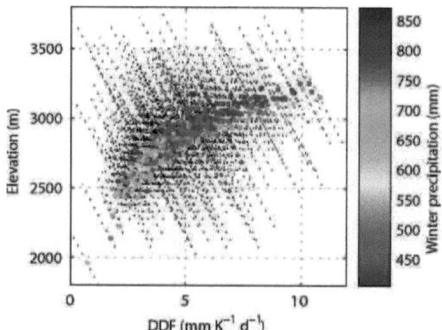

Fig. 5. DDF as a function of elevation for each glacier (thin dashed lines). The colour of the dots shows the mean winter precipitation at the glacier's coordinate according to HISTALP (October–May, colour scale); their location is determined by the DDF at the glacier's median elevation (left and bottom scale).

smaller MD_{abs}, MD and STD, and larger R. The MD values, small in all cases but largest for KWF (0.17 m w.e.), suggest that errors largely compensate each other. The correlation coefficients between measured and modelled balance are similar for the four glaciers with directly measured data: they are highest for HEF (0.85) and lowest for VF (0.75) and on average 0.81. The performance of the model using a linear fit versus a second-order polynomial fit is compared: on average the second-order polynomial fit produces slightly higher correlation coefficients, which is why it is used for the final result.

The results of run II are shown in Figure 4 for all Austrian glaciers for which two DEMs exist (761 out of 900 glaciers; 451 out of 470 km², i.e. 96% of the total glacier area). The evolution of modelled cumulative annual b with time is shown (each black line corresponds to one glacier). In total, 9.4 m w.e. on the ice-covered area of 451 km² was lost between 1969 and 1998, which amounts to 4.7 km³ w.e. The wide spread of the curves is due to differences in glacier type and to the individual glacier's deviation from steady state. The red curves show the cumulative modelled b of the glaciers with directly measured mass balance, and the total, area-weighted cumulative mean specific mass balance is computed and displayed as a green line. Initially, the glaciers' masses remained constant. Starting in 1975, all glaciers show a period of positive mass balances, with a total gain of ~1.3 m w.e., corresponding to an ice volume gain of ~1 km³. Since 1980, only a few years (e.g. 1984 and 1989) interrupt the generally negative trend. This temporal sequence is in agreement with the cumulative volume change derived from direct surface mass-balance measurements as compiled, for example, in Abermann and others (2009).

In Figure 5 the set of DDFs that best reproduced the observed glacier mass loss, the glacier's median elevation (location of the circles) and the mean winter precipitation at the glacier (colours) are displayed. Each thin line represents the DDF(z) of the run with temporally constant DDFs. There is a clear relationship between the median elevation and the DDF at this elevation. The higher the median elevation, the more melt generally occurs at a given temperature and given precipitation. There are two possible explanations for this apparent paradox. First, glaciers usually exist at low median elevations (below 2700 m a.s.l.) because they generally accumulate more mass in winter than glaciers with high median elevations do. This is evident in Figure 5, where glaciers with low median elevations show greater precipitation values (blue to yellow colours). However, more wind drift, more avalanching or some combination of these factors can give an additional mass input. As a result, ice is kept under snow cover for a longer part of the ablation season, so the albedo is comparatively high. Second, increasing shortwave radiation with elevation due to less shading and fewer aerosols results in higher general DDF values at higher elevations, as suggested by Lang and Braun (1990) and Hock (2003).

6. DISCUSSION

The dependence of DDF with time is a major challenge when attempting to model future glacier change. For example, consider the problem faced by a hypothetical modeller in 1975 who sought to model future glacier mass balance through 1998 using mass-balance records from Figure 2 since 1969. He could not have known the future time dependence of DDF, and any reasonable assumption would have resulted in a significant underestimation of ice loss, as is obvious from Figure 2.

The subject of temporally changing DDFs was addressed by Huss and others (2009) who found a decrease in DDFs of ~7% per decade between 1970 and 2008. This seems to contradict the findings of this study but can be explained by the fact that Huss and others (2009) calculated point mass balance, and thus DDFs at point locations that are all in the accumulation area. The increase in DDFs in our study amounts to ~9% per decade and must be understood as a collective signal for an entire glacier. Since mass balances in recent years have been governed by ablation, it is the ablation area that controls recent glacier-averaged DDF changes. A probable explanation of the glacier-averaged DDF increase could be the increased duration of bare-ice exposure and thus a lower mean surface albedo in later years of the study period. Furthermore, two recent studies show that ice albedo recently decreased significantly as a result of additional dust accumulation after a sequence of negative mass-balance years (Oerlemans and others, 2009; Fischer, 2010). The recent accelerated glacier retreat may be partly due to this.

An additional explanation for the temporal increase of the DDFs could be found by considering changes of glacier flow velocities in recent years. Less mass accumulated in upper parts of a glacier leads to less mass turnover and thus less ice supply for the glacier tongues. A strong velocity decrease has been found by exploiting long-term velocity measurements on HEF (Span and others, 1997) and KWF (Abermann and others, 2007). Which of these three processes (albedo, decreased mass transport and dynamical thinning) is the strongest component is beyond the scope of this study. A more process-based approach to separate components of the energy balance, combined with a dynamical ice-flow model, is needed.

This study has several simplifications: The fraction of volume change that is a result of basal melt is not accounted for. This volume loss is reflected in measured volume changes but not in the modelled surface balance as too little is known about its magnitude to accurately model it. Locally, this may be an important contribution to a glacier's mass loss, but on average, and compared to the large surface mass loss, basal melt can be assumed to be negligible (Hooke, 2005).

Another simplification is the assumption that all volume was lost as ice. Several aerial photographs were examined qualitatively, indicating that a fraction of the lost volume was snow. For a study of this scale, however, it is not possible to attribute a proportion of the lost volume that was lost as snow to each glacier. Therefore, to estimate the possible impact on the results, the model was run with the exaggerated assumption that half of the lost volume was snow with a density of 550 kg m^{-3}. The resulting total volume loss then amounts to 3.87 km^3 w.e. or 9% less than calculated with ice density only (i.e. 4.74 km^3 w.e. loss). The pattern of the cumulative mass balance as well as the modifications of DDF as a function of time and space do not change considerably, so the results would be altered quantitatively but not qualitatively.

The sensitivity of the tuning parameters to the result of this study was investigated by altering them stepwise and comparing the difference between the altered and the best-fitting (run II) mean annual balances. If the DDF is increased systematically by 1 mm K^{-1} d^{-1}, mean annual balances for all modelled glaciers become more negative by 600 mm w.e. Likewise, the sensitivity of the vertical gradient of the DDFs (i.e. steepening or flattening the slope of the black lines in Fig. 5) is investigated. Model results are not sensitive to changes in DDF gradients: Changing the vertical dependence of DDF from −0.32 mm^{-1}(Kd)$^{-1}$(100 m)$^{-1}$ to −0.20 mm (Kd)$^{-1}$(100 m)$^{-1}$ (i.e. steepening the slope with zero change at the glacier's median elevation) leads to a reduction of mean annual b by 4 mm. Flattening the slope with zero change at the glacier's median elevation (i.e. −0.44 mm (Kd)$^{-1}$ (100 m)$^{-1}$) leads to a reduction of mean annual b by 9 mm.

The data presented in this study allow for an estimation of the relative importance of the main climatic parameters that govern annual glacier mass balances. In Table 3 we

Table 3. Correlation coefficients of the four variables' anomalies (winter (October–May) (δP_{winter}) and summer (June–August) precipitation (δP_{summer}) and annual ($\delta \Sigma PDD_{year}$) and summer ($\delta \Sigma PDD_{summer}$) PDD sums) with measured balances. Bold numbers indicate statistical significance at the 99% confidence level

	HEF	VF	KWF	SSK
δP_{winter}	0.36	0.18	0.07	0.12
δP_{summer}	0.11	0.20	0.13	0.39
$\delta \Sigma PDD_{year}$	**−0.66**	**−0.66**	**−0.59**	**−0.71**
$\delta \Sigma PDD_{summer}$	**−0.72**	**−0.69**	**−0.74**	**−0.78**

summarize correlation coefficients between measured balances of the four well-studied glaciers and anomalies of PDD sums for the whole year, PDD sums for the summer (June–August), winter precipitation (October–May) and summer precipitation, all of which are interpolated to the glacier's gridpoint. Deviations of winter and summer precipitation correlate only weakly with the measured balance. Annual anomalies of PDD sums correlate much more strongly with surface mass balance than precipitation anomalies do. The strongest correlation between meteorological data and measured balances is found between summer anomalies of PDD sums and the measured balance b (between −0.71 and −0.74), which is consistent with the results of Kuhn and others (1999), who examined this for HEF. All four glaciers have similar correlation values, but SSK shows a significantly higher correlation between summer precipitation anomalies and b than the other glaciers, and HEF's balance is more sensitive than those of the other glaciers to winter precipitation.

The results presented in Figure 4 can be taken as an indication of the generalizability of measured mass balances to all Austrian glaciers modelled in this study. HEF has a much more negative mass balance than most other glaciers, so its b is not representative for the overall mass balance of all Austrian glaciers. The specific balances of VF and KWF are close to the average, while SSK's mean specific balance is more positive than the average.

7. CONCLUSIONS

The change over time in the mean annual surface mass balance of all Austrian glaciers for which two DEMs exist between 1969 and 1998 has been investigated in this study. With a PDD model with DDFs that vary in space (3-D) and time, correlation coefficients of >0.8 on average between measured and modelled balances for the four glaciers with direct measurements for validation are reached. Tuning parameters have to be calibrated to each glacier individually to reach a satisfying result. A relation is found between the glacier's median elevation, mean winter precipitation and the DDF. The general increase in DDFs with time is likely to be due to a larger fraction of the glacier surface that exposes bare ice for an increasingly longer part of the summer, and thus a lower mean albedo. This is consistent with a sequence of negative balance years and an increase in equilibrium-line altitude. Calibrating a PDD model with only a few years of measurements and then extrapolating to the past or future seems extremely unreliable without additional knowledge (e.g. DEMs at various points in time). There are ongoing efforts to extend this study further towards the present and include a new glacier inventory of 2006 (Abermann and others, 2009) in order to investigate how these parameters have evolved in the very recent past. This model could also be used to reproduce mass balances back to the Little Ice Age glacier maximum, for which information on extent and surface elevation could be reconstructed from locations of moraines, thus providing a fascinating picture of >150 years of glacier history.

ACKNOWLEDGEMENTS

This study was funded by the Commission for Geophysical Research, Austrian Academy of Sciences. We thank the Institute of Meteorology and Geophysics, Innsbruck, the Bavarian Academy of Sciences, Munich, Germany, and H. Slupetzky for providing mass-balance data. The ECMWF/ERA-40 reanalysis project and the HISTALP database provided the meteorological data. We thank E. Dreiseitl, E. Schlosser, L.A. Rasmussen and S. Kinter for comments and proofreading. Two anonymous reviewers and the editor are acknowledged for helpful suggestions.

REFERENCES

Abermann, J., H. Schneider and A. Lambrecht. 2007. Analysis of surface elevation changes on Kesselwand glacier – comparison of different methods. *Z. Gletscherkd. Glazialgeol.*, **41**, 147–167.

Abermann, J., A. Lambrecht, A. Fischer and M. Kuhn. 2009. Quantifying changes and trends in the glacier area and volume in the Austrian Ötztal Alps (1969–1997–2006). *Cryosphere*, **3**(2), 205–215.

Abermann, J., M. Kuhn and A. Fischer. 2011. Climatic controls of glacier distribution and glacier changes in Austria. *Ann. Glaciol.*, **52**(59) (see paper in this issue).

Auer, I. *and 23 others*. 2005. A new instrumental precipitation dataset for the greater alpine region for the period 1800–2002. *Int. J. Climatol.*, **25**(2), 139–166.

Böhm, R., I. Auer, M. Granekind and A. Orlik. 2008. Zwei Jahrhunderte Klimaschwankungen in zwei Tälern der Zentralalpen. *Jahrb. Sonnblick-Ver.*

Braithwaite, R.J. 2002. Glacier mass balance: the first 50 years of international monitoring. *Progr. Phys. Geogr.*, **26**(1), 76–95.

Dyurgerov, M.B. and M.F. Meier. 2005. *Glaciers and the changing Earth system: a 2004 snapshot*. Boulder, CO, University of Colorado. Institute of Arctic and Alpine Research. (INSTAAR Occasional Paper 58.)

Efthymiadis, D. *and 7 others*. 2006. Construction of a 10-min-gridded precipitation data set for the Greater Alpine Region for 1800–2003. *J. Geophys. Res.*, **111**(D1), D01105. (10.1029/2005JD006120.)

Escher-Vetter, H., M. Kuhn and M. Weber. 2009. Four decades of winter mass balance of Vernagtferner and Hintereisferner, Austria: methodology and results. *Ann. Glaciol.*, **50**(50), 87–95.

Fischer, A. 2010. Glaciers and climate change: interpretation of 50 years of direct mass balance of Hintereisferner. *Global Planet. Change*, **71**(1–2), 13–26.

Fischer, A. and G. Markl. 2008. Mass balance measurements on Hintereisferner, Kesselwandferner and Jamtalferner 2003 to 2006. Database and results. *Z. Gletscherkd. Glazialgeol.*, **42**(1), 47–83.

Fountain, A.G., M.J. Hoffman, F. Granshaw and J. Riedel. 2009. The 'benchmark glacier' concept – does it work? Lessons from the North Cascade Range, USA. *Ann. Glaciol.*, **50**(50), 163–168.

Frei, C. and C. Schär. 1998. A precipitation climatology of the Alps from high-resolution rain-gauge observations. *Int. J. Climatol.*, **18**(8), 873–900.

Greuell, J.W. and T. Konzelmann. 1994. Numerical modeling of the energy balance and the englacial temperature of the Greenland ice sheet: calculations for the ETH-Camp location (West Greenland, 1155 m a.s.l.). *Global Planet. Change*, **9**(1–2), 91–114.

Gross, G. 1987. Der Flächenverlust der Gletscher in Österreich 1850–1920–1969. *Z. Gletscherkd. Glazialgeol.*, **23**(2), 131–141.

Hock, R. 2003. Temperature index melt modelling in mountain areas. *J. Hydrol.*, **282**(1–4), 104–115.

Hock, R. 2005. Glacier melt: a review on processes and their modelling. *Progr. Phys. Geogr.*, **29**(3), 362–391.

Hock, R., M. de Woul and V. Radić. 2009. Mountain glaciers and ice caps around Antarctica make a large sea-level rise contribution. *Geophys. Res. Lett.*, **36**(7), L07501. (10.1029/2008GL037020.)

Hoinkes, H. and H. Lang. 1962. Der Massenhaushalt von Hintereis- und Kesselwandferner (Ötztaler Alpen), 1957/58 und 1958/59. *Arch. Meteorol. Geophys. Bioklimatol., Ser. B*, **12**(1), 284–320.

Hooke, R.LeB. 2005. *Principles of glacier mechanics. Second edition*. Cambridge, etc., Cambridge University Press.

Huss, M., A. Bauder, M. Funk and R. Hock. 2008. Determination of the seasonal mass balance of four Alpine glaciers since 1865. *J. Geophys. Res.*, **113**(F1), F01015. (10.1029/2007JF000803.)

Huss, M., M. Funk and A. Ohmura. 2009. Strong Alpine glacier melt in the 1940s due to enhanced solar radiation. *Geophys. Res. Lett.*, **36**(23), L23501. (10.1029/2009GL040798.)

Kaser, G., J.G. Cogley, M.B. Dyurgerov, M.F. Meier and A. Ohmura. 2006. Mass balance of glaciers and ice caps: consensus estimates for 1961–2004. *Geophys. Res. Lett.*, **33**(19), L19501. (10.1029/2006GL027511.)

Kuhn, M. 2003. Redistribution of snow and glacier mass balance from a hydrometeorological model. *J. Hydrol.*, **282**(1–4), 95–103.

Kuhn, M., U. Nickus and F. Pellet. 1982. Die Niederschlagsverhältnisse im inneren Ötztal. In *17. Internationale Tagung für Alpine Meteorologie, 21.–25. September 1982, Berchtesgaden*. Offenbach am Main, Deutscher Wetterdienst, 235–237.

Kuhn, M., G. Markl, G. Kaser, U. Nickus, F. Obleitner and H. Schneider. 1985. Fluctuations of climate and mass balance: different responses of two adjacent glaciers. *Z. Gletscherkd. Glazialgeol.*, **21**(1–2), 409–416.

Kuhn, M., E. Dreiseitl, S. Hofinger, G. Markl, N. Span and G. Kaser. 1999. Measurements and models of the mass balance of Hintereisferner. *Geogr. Ann.*, **81A**(4), 659–670.

Kuhn, M., A. Lambrecht, J. Abermann, G. Patzelt and G. Gross. 2009. *Die österreichischen Gletscher 1998 und 1969, Flächen- und Volumenänderungen*. Wien, Österreichische Akademie der Wissenschaften. (Landesverteidigungsakademie, Bundesministerium für Landesverteidigung Projektbericht 10.)

Lambrecht, A. and M. Kuhn. 2007. Glacier changes in the Austrian Alps during the last three decades, derived from the new Austrian glacier inventory. *Ann. Glaciol.*, **46**, 177–184.

Lang, H. and L. Braun. 1990. On the information content of air temperature in the context of snow melt estimation. *IAHS Publ.* 190 (Symposium at Strbské Pleso 1988 – *Hydrology of Mountainous Areas*), 347–354.

Lemke, P. *and 10 others*. 2007. Observations: changes in snow, ice and frozen ground. *In* Solomon, S. *and 7 others*, eds. *Climate change 2007: the physical science basis. Contribution of Working Group I to the Fourth Assessment Report of the Intergovernmental Panel on Climate Change*. Cambridge, etc., Cambridge University Press, 339–383.

Machguth, H., F. Paul, S. Kotlarski and M. Hoelzle. 2009. Calculating distributed glacier mass balance for the Swiss Alps from regional climate model output: a methodical description and interpretation of the results. *J. Geophys. Res.*, **114**(D19), D19106. (10.1029/2009JD011775.)

Meier, M.F. *and 7 others*. 2007. Glaciers dominate eustatic sea-level rise in the 21st century. *Science*, **317**(5841), 1064–1067.

Oerlemans, J., R.H. Giesen and M.R. van den Broeke. 2009. Retreating alpine glaciers: increased melt rates due to accumulation of dust (Vadret da Morterastch, Switzerland). *J. Glaciol.*, **55**(192), 729–736.

Radić, V. and R. Hock. 2006. Modeling future glacier mass balance and volume changes using ERA-40 reanalysis and climate models: sensitivity study at Storglaciären, Sweden. *J. Geophys. Res.*, **111**(F3), F03003. (10.1029/2005JF000440.)

Rasmussen, L.A. and J.M. Wenger. 2009. Upper-air model of summer balance on Mount Rainier, USA. *J. Glaciol.*, **55**(192), 619–624.

Reinwarth, O. and H. Escher-Vetter. 1999. Mass balance of Vernagtferner, Austria, from 1964/65 to 1996/97: results for three sections and the entire glacier. *Geogr. Ann.*, **81A**(4), 743–751.

Schöner, W. and R. Böhm. 2007. A statistical mass-balance model for reconstruction of LIA ice mass for glaciers in the European Alps. *Ann. Glaciol.*, **46**, 161–169.

Schrott, D. 2006. Flächenhafte Modellierung der Energie- und Massenbilanz am Hintereisferner. (Diplomarbeit, Universität Innsbruck.)

Slupetzky, H. 1989. Die Massenbilanzmessreihe vom Stubacher Sonnblickkees 1958/59 bis 1987/88 – Die Berechnung der Massenbilanz 1980/81 bis 1987/88 und 1958/59 bis 1962/63. *Z. Gletscherkd. Glazialgeol.*, **25**(1), 69–89.

Slupetzky, H. 2003. Do we need long term terrestrial glacier mass balance monitoring for the future? *Geophys. Res. Abstr.*, **5**, EAE03-A-11390.

Span, N., M. Kuhn and H. Schneider. 1997. 100 years of ice dynamics of Hintereisferner, central Alps, Austria, 1894–1994. *Ann. Glaciol.*, **24**, 297–302.

Steinacker, R. 1983. Diagnose und Prognose der Schneefallgrenze. *Wetter Leben*, **35**, 81–90.

Stull, R. 2000. *Meteorology for scientists and engineers. Second edition.* Forest Grove, CA, Brooks/Cole.

Uppala, S.M. *and 45 others*. 2005. The ERA-40 re-analysis. *Q. J. R. Meteorol. Soc.*, **131**(612), 2961–3012.

World Glacier Monitoring Service (WGMS). 2007. *Glacier Mass Balance Bulletin No. 9 (2004–2005),* ed. Haeberli, W., M. Zemp and M. Hoelzle. ICSU (FAGS)/IUGG (IACS)/UNEP/UNESCO/WMO, World Glacier Monitoring Service, Zürich.

APPENDIX B: GLACIER CHANGES - DOCUMENTARY WORK

Historical records can be used to document glacier changes, which has a high potential for bringing environmental issues into public attention. Old photos or paintings are used to determine past glacier extents or find evidence for advance chronologies (Nicolussi, 1993, Nussbaumer and others, 2007). The quality and scientific value depends on the detailedness and accuracy of the painting. It has started to become a hobby of the author to collect such repeat documentations. So far, they have not been exploited quantitatively and there are many more to continue but some first examples are presented here.

The first example (Figure B1) is in the Zillertal Alps and shows Waxeggkees, Großer Möseler and Berliner Hütte painted by the English-German painter and mountaineer E.T. Compton.

Figure B1: Waxeggkees and Berlinerhuette, Zillertal Alps, Austria as painted by E.T. Compton (Brandes and Brandes, 2007) probably around 1900 and photographed in August 2009 (lower part; photo: J. Abermann).

Another example is presented in Figure B2. It is an early painting by F. Thurner from 1938, a painter who mainly spent his life in Innsbruck, Austria. It shows Tyrol's largest glacier, Gepatschferner in Kaunertal. Compared with the example from E.T. Compton, it is slightly more abstract and the author suspects that it was made from several nearby locations as the exact perspective could not be found. This is a common practice of landscape painters to compile their motive from several perspectives and limits the possibility for scientific exploitation. It is obvious that Gepatschferner retreated significantly less than most other glaciers in the 20th century, which is a consequence of the large accumulation-area ratio.

Figure B2: Gepatschferner as painted by F. Thurner in 1938 (upper part), and photographed on 23/07/ 2009 (lower part, photo: J. Abermann).

The next example shows a photographic time series of Kesselwandferner, Ötztal Alps between 1940 and today with varying time-steps in between (Figure B3). It is a unique and highly interesting example as it documents well the advance starting in the early 1970s. Note the heavily crevassed tongue during the advance (e.g. in 1979) and, although at a comparable stage, the smooth surface in 1991 during the recession. The comparison of 1971, 1975 and 1978 can be taken as a visualization for the advection term in the equation of ice motion: Thickening in upper parts of the glacier that is finally transported downwards leading to an advance. The accelerated retreat after 1997 is striking and unless some favourable balance years happen very soon, the glacier will not be visible from this position after few years. It can be suspected, however, that Kesselwandferner will be one of the few glaciers that will reach a state close to equilibrium at least for some years. This will happen right after disappearance from this point as it then fills a plateau at altitudes above 3000 m.

Figure B3: Time series of Kesselwandferner including the advance of the late 1970s and early 1980s. The original pictures are from the archives of H. Schneider, N. Span, the Institute of Meteorology and Geophysics, University of Innsbruck and J. Abermann. Thanks to N. Span for scanning older pictures.

The last example shown is a particularly ideal one because of its artistic quality, its detailedness, the interesting point in time and the large spatial coverage: Figure B4 shows a panorama from Kreuzspitze, Ötztal Alps showing well-studied glaciers like Vernagtferner and Hintereisferner. It was painted in 1869 by C. Jordan and G. Engelhardt on behalf of the founder of the German and Austrian Alpine club, F. Senn (Woebcken, 1989). It displays a unique record of the glacier extent close to the LIA-Maximum. Comparisons were reproduced earlier already by (Woebcken, 1989) and the commission for glaciology from the Bavarian Academy of Sciences (http://www.lrz.de/~a2901ad/webserver/webdata/download/Faltblatt_KfG2006.pdf). The comparison-shot shown here was made on 23th September 2010, the last day before the first permanent snow of this winter fell.

It is interesting to get the dramatic glacier recession visualized in such a detailed way. Whereas the glacier tongues of the large valley glaciers retreated nearly into non-existence, compared to the late 19th century (e.g. Hintereisferner, Schalfferner), the high-plateau of Kesselwandferner or the upper part of Vernagtferner show fairly little differences.

A very striking detail is the thick firn patch that is visible at the painting on the far left and far right side with a probably several meters thick cornice. It entirely disappeared until today. It cannot be excluded that this was painted using 'artistic freedom' - however, a comparison with the same source as shown in Figure 14 supports a very strong recession at that ridge.

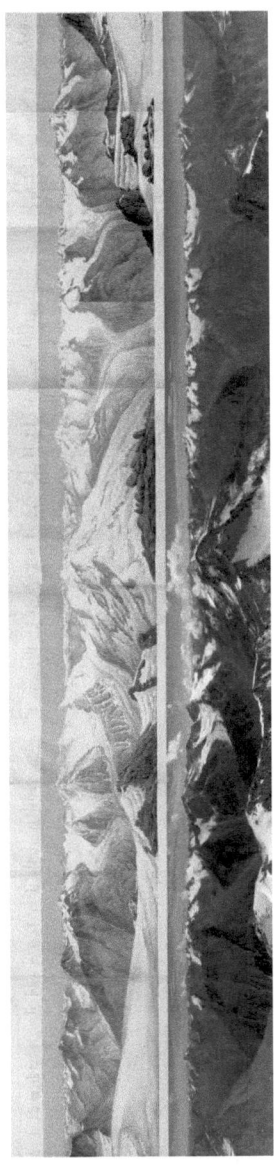

Figure B4: The Panorama of Kreuzspitze as painted in 1869 by C. Jordan and G. Engelhardt (upper part). The digital processing of the original was done by N. Span. The original is at University Library Innsbruck. The comparison panorama was made on 23/9/2010 (lower part, photo: J. Abermann).

REFERENCES

Abermann, J., C. Knoll, H. Kerschner and A. Lambrecht, 2009a. Synchroneous glacier retreat North and South of a Central-Alpine mountain divide? *Poster presentation: Alpine Glaciology Meeting*, Innsbruck, Austria.

Abermann, J., A. Lambrecht, A. Fischer and M. Kuhn, 2009b. Quantifying changes and trends in glacier area and volume in the Austrian Ötztal Alps (1969-1997-2006). *The Cryosphere*, **3**(2): 205-215.

Abermann, J., A. Lambrecht and H. Schneider. 2007. Analysis of surface elevation changes on Kesselwand glacier - Comparison of different methods. *Zeitschrift für Gletscherkunde und Glazialgeologie*, **41**, 147 - 168.

Abermann, J., A. Fischer, A. Lambrecht and T. Geist, 2010a. On the potential of very high-resolution repeat DEMs in glacial and periglacial environments. *The Cryosphere*, **4**(1): 53-65.

Abermann, J., B. Seiser and A. Fischer, 2010b. Towards a third Austrian glacier inventory: First results and a climatic interpretation. *Poster presentation: European Geoscience's General Assembly*, Vienna, Austria.

Abermann, J., M. Kuhn and A. Fischer, 2011a. A reconstruction of annual mass balances of Austria's glaciers from 1969 to 1998. *Annals of Glaciology, 59*(51), 127-134.

Abermann, J., M. Kuhn and A. Fischer, 2011b. Climatic controls of glacier distribution and glacier changes in Austria. *Annals of Glaciology, 59*(50), 83-90.

Andreassen, L.M., F. Paul, A. Kääb and J.E. Hausberg. 2008. Landsat-derived glacier inventory for Jotunheimen, Norway, and deduced glacier changes since the 1930s. *The Cryosphere*, **2**(2): 131-145.

Auer, I., R. Böhm, A. Jurkovic, A. Orlik, R. Potzmann, W. Schöner, M. Ungersböck, M. Brunetti, T. Nanni, M. Maugeri, K. Briffa, P. Jones, D. Efthymiadis, O. Mestre, J.M. Moisselin, M. Begert, R. Brazdil, O. Bochnicek, T. Cegnar, M. Gajic-Capka, K. Zaninovic, Z. Majstorovic, S. Szalai and T. Szentimrey, 2005. A new instrumental precipitation data set in the greater alpine region for the period 1800–2002. *International Journal of Climatology*, **25**(2): 139-166.

Baker, D., H. Escher-Vetter, H. Moser, H. Oerter and O. Reinwarth, 1982. A glacier discharge model based on results from field studies of energy balance, water storage and flow. *IAHS Publ.*, **138**: 103-112.

Bamber, J.L. and A. Rivera, 2007. A review of remote sensing methods for glacier mass balance determination. *Global and Planetary Change*, **59**(1-4): 138-148.

Barker, P.F., B. Diekmann and C. Escutia, 2007. Onset of Cenozoic Antarctic glaciation. *Deep Sea Research Part II: Topical Studies in Oceanography*, **54**(21-22): 2293-2307.

Barry, R.G., 1992. *Mountain weather and climate*, Routledge, London, 422 pp.

Beer, J., W. Mende and R. Stellmacher, 2000. The role of the sun in climate forcing. *Quaternary Science Reviews*, **19**(1-5): 403-415.

Benn, D. and D.J.A. Evans, 2010. *Glaciers and Glaciation.* 2 ed., Hodder Education, Oxon, UK, 802pp.

Böhm, R., I. Auer, M. Ganekind and A. Orlik. 2008. Zwei Jahrhunderte Klimaschwankungen in zwei Tälern der Zentralalpen. *Jahresbericht des Sonnblickvereins*, **103-104**: 10-20.

Bolch, T., B. Menounos and R. Wheate. 2010. Landsat-based inventory of glaciers in western

Canada, 1985-2005. *Remote Sensing of Environment*, **114**(1): 127-137.

Bonani, G., S. Ivy, I. Hajdas, T.R. Niklaus and M. Suter, 1994. AMS 14C age determinations of tissue, bone and grass samples from the Ötztal Ice Man. *Radiocarbon*, **36**: 247-250.

Bortenschlager, S., 1984. Beiträge zur Vegetationsgeschichte Tirols I. Inneres Ötztal und unteres Inntal. *Berichte des naturwissenschaftlich-medizinischen Vereins Innsbruck*, **71**: 19-56.

Bradley, R.S., 1985. *Quaternary Paleoclimatology*. Allen&Unwin, London, UK, 472pp.

Bradley, R.S., 1999. *Paleoclimatology: Reconstructing Climates of the Quarternary*. 2 ed., Academic Press, San Diego, USA, 613pp.

Brandes, J. and S. Brandes, 2007. E. T. Compton. Bergverlag Rother, Munich, Germany, 392pp.

Braithwaite, R. and Y. Zhang, 2000. Sensitivity of mass balance of five Swiss glaciers to temperature changes assessed by tuning a degree-day model. *Journal of Glaciology*, **46**(152): 7-14.

Braithwaite, R.J., 2002. Glacier mass balance: the first 50 years of international monitoring. *Progress in Physical Geography*, **26**(1): 76-95.

Braun, L.N., M. Weber and M. Schulz, 2000. Consequences of climate change for runoff from Alpine regions. *Annals of Glaciology*, **31**: 19-25.

Collins, D.N., 1984. Water and Mass Balance Measurements in Glacierised Drainage Basins. *Geografiska Annaler. Series A, Physical Geography*, **66**(3): 197-214.

Croll, J., 1867a. On the change in the obliquity of the eliptic, its influence on the climate of the polar regions and on the level of the sea. *Philosophical Magazine*, **33**: 426-445.

Croll, J., 1867b. On the excentricity of the Earth's orbit, and its physical relations to the glacial epoch. *Philosophical Magazine*, **33**: 119-131.

Crowell, J.C., 1999. Pre-Mesozoic Ice Ages: Their Bearing on Understanding the Climate System. *Geological Society of America Memoirs*, **192**: 1-112.

Cuffey, K.M. and W.S.B. Paterson, 2010. *The physics of glaciers.* 4 ed., Butterworth-Heinemann, Burlington, US, 650 pp.

DeBeer, C.M. and M.J. Sharp. 2009. Topographic influences on recent changes of very small glaciers in the Monashee Mountains, British Columbia, Canada. *Journal of Glaciology*, **55**(192): 691-700.

Deline, P. and G. Orombelli, 2005. Glacier fluctuations in the western Alps during the Neoglacial as indicated by the Miage morainic amphitheatre (Mont Blanc massif, Italy). *Boreas*, **34**: 1-12.

Dyurgerov, M. and M.F. Meier, 2005. Glaciers and Changing Earth System: A 2004 Snapshot. INSTAAR 58, Boulder, USA, 117 pp.

Eder, K., R. Würländer and H. Rentsch, 2000. Digital photogrammetry for the new glacier inventory of Austria. *IAPRS International Archives of Photogrammetry and Remote Sensing*, **33**: 1-15.

Ehlers, J. and P.L. Gibbard, 2007. The extent and chronology of Cenozoic Global Glaciation. *Quaternary International*, **164-165**: 6-20.

Efthymiadis, D., P.D. Jones, K.R. Briffa, I. Auer, R. Böhm, W. Schöner, C. Frei and J. Schmidli. 2006. Construction of a 10-min-gridded precipitation data set for the Greater Alpine Region for 1800–2003. *Journal of Geophysical Research*, **111**(D01105), 1-22.

Escher-Vetter, H., M. Kuhn and M. Weber, 2009. Four decades of winter mass balance of Vernagtferner and Hintereisferner, Austria: methodology and results. *Annals of Glaciology*, **50**: 87-

95.

Evans, D.A., N.J. Beukes and J.L. Kirschvink, 1997. Low-latitude glaciation in the Palaeoproterozoic era. *Nature*, **386**(6622): 262-266.

Evans, I.S. 2006. Glacier Distribution in the Alps: Statistical Modelling of Altitude and Aspect. *Geografiska Annaler: Series A, Physical Geography*, **88**(2), 115-133.

Evans, I.S. and N.J. Cox. 2005. Global variations of local asymmetry in glacier altitude: separation of northsouth and eastwest components. *Journal of Glaciology*, **51**(174), 469-482.

Evans, I.S. and N.J. Cox. 2010. Climatogenic northsouth asymmetry of local glaciers in Spitsbergen and other parts of the Arctic. *Annals of Glaciology*, **51**(192), 16-22.

Finsterwalder, S. and H. Schunk, 1887. Der Suldenferner. *Zeitschrift des Deutschen und Österreichischen Alpenvereins*, **18**: 72-89.

Fischer, A., 2010. Glaciers and climate change: Interpretation of 50 years of direct mass balance of Hintereisferner. *Global and Planetary Change*, **71**(1-2): 13-26.

Fischer, A. and G. Markl. 2008. Mass balance measurements on Hintereisferner, Kesselwandferner, and Jamtalferner 2003 to 2006. Database and results. *Zeitschrift für Gletscherkunde und Glazialgeologie*, **42**(1), 47-83.

Fischer, A., M. Olefs and J. Abermann, 2010. Glaciers, snow and ski tourism in Austria's changing climate. *accepted for Annals of Glaciology*.

Fliri, F. 1975. *Das Klima im Raume von Tirol*. Innsbruck-München, Universitätsverlag Wagner.

Fountain, A.G., M.J. Hoffman, F. Granshaw and J. Riedel. 2009. The 'benchmark glacier' concept does it work? Lessons from the North Cascade Range, USA. *Annals of Glaciology*, **50**, 163-168.

Fraedrich, R., 1975. Spät- und postglaziale Gletscherschwankungen in der Ferwallgruppe (Tirol/Vorarlberg). *Düsseldorfer Geographische Schriften*, **12**: 1-161.

Free, M. and A. Robock, 1999. Global warming in the context of the Little Ice Age. *J. Geophys. Res.*, **104**(D16): 19057-19070.

Frei, C. and C. Schär. 1998. A precipitation climatology of the Alps from high-resolution raingauge observations. *International Journal of Climatology*, **18**(8), 873-900.

Goller, M., 2010. *Gletscherinventar von Vorarlberg und Westtirol von 2006*. Bachelor's Thesis, University of Innsbruck, 45 pp.

Greuell, W. and T. Konzelmann. 1994. Numerical modelling of the energy balance and the englacial temperature of the Greenland Ice Sheet. Calculations for the ETH-Camp location (West Greenland, 1155 m a.s.l.). *Global and Planetary Change*, **9**(1-2), 91-114.

Gross, G., 1987. Der Flächenverlust der Gletscher in Österreich 1850–1920–1969. *Zeitschrift für Gletscherkunde und Glazialgeologie*, **23**(2): 131-141.

Gross, G., H. Kerschner and G. Patzelt, 1978. Methodische Untersuchungen über die Schneegrenze in alpinen Gletschergebieten. *Zeitschrift für Gletscherkunde und Glazialgeologie*, **12**(2): 223-251.

Grotzinger, J., T.H. Jordan, F. Press and R. Siever, 2007. *Understanding Earth*. 5 ed., Freeman, New York, USA, 672 pp.

Grove, J.M., 2004a. *Little ice ages : ancient and modern. Volume I*, Routledge, New York, USA, 402pp.

Grove, J.M., 2004b. *Little ice ages : ancient and modern. Volume II*, Routledge, New York, USA, 316pp.

Grove, J.M. and R. Switsur, 1994. Glacial geological evidence for the Medieval Warm Period. *Climate Change*, **26**: 143-169.

Haeberli, W. and M. Hoelzle. 1995. Application of inventory data for estimating characteristics of and regional climate-change effects on mountain glaciers: a pilot study with the European Alps. *Annals of Glaciology*, **21**: 206-212.

Hays, J.D., J. Imbrie and N.J. Shackleton, 1976. Variations in the Earth's Orbit: Pacemaker of the Ice Ages. *Science*, **194**(4270): 1121-1132.

Heuberger, H., 1966. Gletschergeschichtliche Untersuchungen in den Zentralalpen zwischen Sellrain und Ötztal. *Wissenschaftliche Alpenvereinshefte*, **20**: 1-126.

Hock, R., 2003. Temperature index melt modelling in mountain areas. *Journal of Hydrology*, **282**(1-4): 104-115.

Hock, R. 2005. Glacier melt: A review on processes and their modelling. *Progress in Physical Geography*, 29(3), 362-391.

Hock, R., V. Radic and M. De Woul, 2007. Climate sensitivity of Storglaciaren, Sweden: an intercomparison of mass-balance models using ERA-40 re-analysis and regional climate model data. *Annals of Glaciology*, **46**(1): 342-348.

Hock, R., M. de Woul, Radic, V. and M. Dyurgerov. 2009. Mountain glaciers and ice caps around Antarctica make a large sea-level rise contribution. *Geophys. Res. Lett.*, **36**(7), L07501.

Hoelzle, M., T. Chinn, D. Stumm, F. Paul, M. Zemp and W. Haeberli. 2007. The application of glacier inventory data for estimating past climate change effects on mountain glaciers: A comparison between the European Alps and the Southern Alps of New Zealand. *Global and Planetary Change*, **56**(1-2): 69-82.

Hoinkes, H. and H. Lang. 1962. Der Massenhaushalt von Hintereis- und Kesselwandferner (Ötztaler Alpen), 1957/58 und 1958/59. *Theoretical and Applied Climatology*, **12**(2), 284-320.

Hoinkes, 1970. Methoden und Möglichkeiten von Massenhaushaltsstudien auf Gletschern. *Zeitschrift für Gletscherkunde und Glazialgeologie*, 6(1-2): 37-90.

Holzhauser, H., M. Magny and H.J. Zumbühl, 2005. Glacier and lake-level variations in west-central Europe over the last 3500 years. *The Holocene*, **15**(6): 789-801.

Hooke, R.L. 2005. *Principles of Glacier Mechanics*. Cambridge, UK, Camebridge University Press, 448pp.

Huss, M. and A. Bauder, 2009. 20th-century climate change inferred from four long-term point observations of seasonal mass balance. *Annals of Glaciology*, **50**: 207-214.

Huss, M., A. Bauder, M. Funk and R. Hock, 2008. Determination of the seasonal mass balance of four Alpine glaciers since 1865. *Journal of Geophysical Research*, **113**(F01015).

Huss, M., M. Funk and A. Ohmura, 2009. Strong Alpine glacier melt in the 1940s due to enhanced solar radiation. *Geophys. Res. Lett.*, **36**(23): L23501.

Ivy-Ochs, S., H. Kerschner, M. Maisch, M. Christl, P.W. Kubik and C. Schlüchter, 2009. Latest Pleistocene and Holocene glacier variations in the European Alps. *Quaternary Science Reviews*, **28**(21-22): 2137-2149.

Ivy-Ochs, S., H. Kerschner, A. Reuther, F. Preusser, K. Heine, M. Maisch, P.W. Kubik and C. Schlüchter, 2008. Chronology of the last glacial cycle in the European Alps. *Journal of Quaternary Science*, **23**(6-7): 559-573.

Ivy-Ochs, S., J. Schäfer, P. Kubik, H.-A. Synal and C. Schlüchter, 2004. Timing of deglaciation on the northern Alpine foreland (Switzerland). *Ecologae Geologicae Helvetiae*, **97**: 47-55.

Joerin, U.E., T.F. Stocker and C. Schlüchter, 2006. Multicentury glacier fluctuations in the Swiss Alps during the Holocene. *The Holocene*, **16**(5): 697-704.

Kääb, A., F. Paul, M. Maisch and W. Häberli, 2002. The new remote-sensing-derived Swiss Glacier Inventory: II. First results. *Annals of Glaciology*, **34**: 362-366.

Kalnay, E., M. Kanamitsu, R. Kistler, W. Collins, D. Deaven, L. Gandin, M. Iredell, S. Saha, G. White, J. Woollen, Y. Zhu, A. Leetmaa, R. Reynolds, M. Chelliah, W. Ebisuzaki, W. Higgins, J. Janowiak, K.C. Mo, C. Ropelewski, J. Wang, R. Jenne and D. Joseph, 1996. The NCEP/NCAR 40-Year Reanalysis Project. *Bulletin of the American Meteorological Society*, **77**(3): 437-471.

Kääb, A., F. Paul, M. Maisch and W. Häberli. 2002. The new remote-sensing-derived Swiss Glacier Inventory: II. First results. *Annals of Glaciology*, **34**: 362-366.

Kaser, G., J.G. Cogley, M.B. Dyurgerov, M.F. Meier and A. Ohmura. 2006. Mass balance of glaciers and ice caps: Consensus estimates for 1961-2004. *Geophys. Res. Lett.*, **33**(19), L19501.

Kaser, G., A. Fountain and P. Jansson, 2003a. A manual for monitoring the mass balance of mountain glaciers with particular attention to low latitude characteristics. IHP-VI 59, UNESCO, Paris, France, 137pp.

Kaser, G., I. Juen, C. Georges, J. Gómez and W. Tamayo, 2003b. The impact of glaciers on the runoff and the reconstruction of mass balance history from hydrological data in the tropical Cordillera Blanca, Peru. *Journal of Hydrology*, **282**: 130-144.

Kerschner, H., A. Hertl, G. Gross, S. Ivy-Ochs and P.W. Kubik, 2006. Surface exposure dating of moraines in the Kromer valley (Silvretta Mountains, Austria)-evidence for glacial response to the 8.2 ka event in the Eastern Alps? *The Holocene*, **16**(1): 7-15.

Kilger, F., 1892. Hochtouren im Mieminger Gebirge. *Zeitschrift des Deutschen und Österreichischen Alpenvereins*, **23**: 84-123.

Klok, E.J. and J. Oerlemans, 2002. Model study of the spatial distribution of the energy and mass balance of Morteratschgletscher, Swtizerland. *Journal of Glaciology*, **48**(163): 505-518.

Knoll, C., 2006. *Gletscherinventar Südtirol AA1997*. Diploma thesis, University of Innsbruck, 103pp.

Knoll, C., H. Kerschner and J. Abermann, 2009. Development of South Tyrolean glaciers since the Little Ice Age maximum. *Zeitschrift für Gletscherkunde und Glazialgeologie*, **42**(1): 19-36.

Kuhn, M., 1993. Der Mieminger Schneeferner, ein Beispiel eines lawinengenährten Kargletschers. *Zeitschrift für Gletscherkunde und Glazialgeologie*, **29**(2): 153-171.

Kuhn, M., 1995. The mass balance of very small glaciers. *Zeitschrift für Gletscherkunde und Glazialgeologie*, **31**: 171 - 179.

Kuhn, M., 2003. Redistribution of snow and glacier mass balance from a hydrometeorological model. *Journal of Hydrology*, **282**: 95-103.

Kuhn, M., (ed.), 2007. *Omega*, Zeitschrift für Gletscherkunde und Glazialgeologie, 41, Universitätsverlag Wagner, Innsbruck, Austria, 232pp.

Kuhn, M., J. Abermann, M. Bacher and M. Olefs. 2008. The transfer of mass-balance profiles to unmeasured glaciers *Annals of Glaciology*, **50**(50), 185-190.

Kuhn, M., E. Dreiseitl, S. Hofinger, G. Kaser, G. Markl and N. Span, 1999. Measurements and Models of the Mass Balance of Hintereisferner. *Geografiska Annaler*, **81A**(4): 659 - 670.

Kuhn, M., A. Lambrecht, J. Abermann, G. Patzelt and G. Gross, 2009. *Die österreichischen Gletscher 1998 und 1969, Flächen und Volumenänderungen.* Austrian Academy of Sciences Press, Vienna, Austria, 125pp.

Kuhn, M., G. Markl, G. Kaser, U. Nickus, F. Obleitner and H. Schneider. 1985. Fluctuations of climate and mass balance: different responses of two adjacent glaciers. *Zeitschrift für Gletscherkunde und Glazialgeologie*, **21**, 409 - 416.

Kuhn, M., U. Nickus and F. Pellet. 1982. Die Niederschlagsverhältnisse im inneren Ötztal. *Internationale Tagung für Alpine Meteorologie*, Berchtesgaden, 235-237.

Lambrecht, A. and M. Kuhn, 2007. Glacier changes in the Austrian Alps during the last three decades, derived from the new Austrian glacier inventory. *Annals of Glaciology*, **46**: 177-184.

Lambrecht, A. and C. Mayer, 2009. Temporal variability of the non-steady contribution from glaciers to water discharge in western Austria. *Journal of Hydrology*, **376**(3-4): 353-361.

Lang, H. and L. Braun, 1990. On the information content of air temperature in the context of snow melt estimation. In Molnar, L., (ed.), *Hydrology of Mountainouos Areas, Proceedings of the Strbske Pleso Symposium 1990*, 347-354.

Lehning, M., H. Löwe, M. Ryser and N. Raderschall, 2008. Inhomogeneous precipitation distribution and snow transport in steep terrain. *Water Resour. Res.*, **44**(7): W07404.

Lemke, P., J. Ren, R.B. Alley, I. Allison, J. Carrasco, G. Flato, Y. Fujii, G. Kaser, P. Mote, R,H. Thomas and T. Zhang, 2007. In Observations: Changes in Snow, Ice and Frozen Ground. In Solomon, S., D. Qin, M. Manning, Z. Chen, M. Marquis, K.B. Averyt, M. Tignor and H.L. Miller, (eds.), *Climate Change 2007: The Physical Science Basis. Contribution of Working Group I to the Fourth Assessment Report of the Intergovernmental Panel on Climate Change*, Cambridge University Press, Cambridge, UK and New York, USA, 1009pp.

Machguth, H., F. Paul, S. Kotlarski and M. Hoelzle, 2009. Calculating distributed glacier mass balance for the Swiss Alps from regional climate model output: A methodical description and interpretation of the results. *J. Geophys. Res.*, **114**(D19): D19106.

Maisch, M., 1981. Glazialmorphologische und gletschergeschichtliche Untersuchungen im Gebiet zwischen Landwasser- und Albulatal (Kt. Graubünden, Schweiz). *Physische Geographie*, **3**: 1-215.

Matthes, F.E., 1939. Report of Committee on Glaciers, April 1939. *Transactions of the American Geophysical Union*, **20**: 518-523.

Matthews, J.A. and K.R. Briffa, 2005. The 'Little Ice Age': Re-evaluation of an evolving concept. *Geografiska Annaler: Series A, Physical Geography*, **87**(1): 17-36.

Mayr, F. and H. Heuberger, 1968. Type Areas of Late Glacial and Post Glacial deposits in Tyrol, Eastern Alps. In Richmond, G.M., (ed.), *Glaciation of the Alps*, University of colorado Studies in Earth Sciences, 143 - 165.

Meier, M.F., M.B. Dyurgerov, U.K. Rick, S. O'Neel, W.T. Pfeffer, R.S. Anderson, S.P. Anderson and A.F. Glazovsky. 2007. Glaciers Dominate Eustatic Sea-Level Rise in the 21st Century. *Science*, **317**(5841), 1064-1067.

Milankovitch, M., 1941. *Kanon der Erdbestrahlung und seine Anwendung auf das Eiszeitenproblem.* Königlich Serbische Akademie, Belgrade, Serbia, 633 pp.

Mölg, T. and D.R. Hardy, 2004. Ablation and associated energy balance of a horizontal glacier surface on Kilimanjaro. *Journal of Geophysical Research*, **109**(D16104, doi:10.1029/2003JD004338): 1-13.

Muttoni, G., C. Carcano, E. Garzanti, M. Ghielmi, A. Piccin, R. Pini, S. Rogledi and D. Sciunnach,

2003. Onset of major Pleistocene glaciations in the Alps. *Geology*, **31**(11): 989-992.

Nesje, A. and S.O. Dahl, 2000. *Glaciers and Environmental Change*. Arnold, London, UK, 203pp.

Nesje, A. and S.O. Dahl, 2003. The 'Little Ice Age' – only temperature? *The Holocene*, **13**(1): 139-145.

Nicolussi, K., 1993. Bilddokumente zur Geschichte des Vernagtferners im 17.Jahrhundert. *Zeitschrift für Gletscherkunde und Glazialgeologie*, **26**(2): 97-119.

Nicolussi, K., M. Kaufmann, G. Patzelt, J. van der Plicht and A. Thurner, 2005. Holocene tree-line variability in the Kauner Valley, Central Eastern Alps, indicated by dendrochronological analysis of living trees and subfossil logs. *Vegetation History and Archaeobotany*, **14**(3): 221-234.

Nicolussi, K. and G. Patzelt, 2001. Untersuchungen zur holozänen Gletscherentwicklung von Pasterze und Gepatschferner (Ostalpen). *Zeitschrift für Gletscherkunde und Glazialgeologie*, **36**: 1-87.

Nussbaumer, S.U., 2010. *Continental-scale glacier variations in Europe (Alps, Scandinavia) and their connection to climate over the last centuries*. PhD-Thesis, University of Bern, 309pp.

Nussbaumer, S.U., H.J. Zumbühl and D. Steiner, 2007. Fluctuations of the Mer de Glace (Mont Blanc area, France) AD 1500-2050. Part I: The history of the Mer de Glace AD 1570-2003 according to pictorial and written documents. *Zeitschrift für Gletscherkunde und Glazialgeologie*, **40**: 5-40.

Oerlemans, J., R.H. Giesen and M.R. Van Den Broeke. 2009. Retreating alpine glaciers: increased melt rates due to accumulation of dust (Vadret da Morteratsch, Switzerland). *Journal of Glaciology*, **55**, 729-736.

Ohmura, A., 2001. Physical basis for the temperature/melt-index method. *Journal of Applied Meteorology*, **40**: 753-761.

Ohmura, A., 2009. Completing the World Glacier Inventory. *Annals of Glaciology*, **50**(53): 144-148.

Orombelli, G. and P. Mason, 1997. Holocene glacier fluctuations in the Italian Alpine region. *Paläoklimaforschung*, **24**: 59-65.

Patzelt, G., 1972. Die spätglazialen Stadien und postglazialen Schwankungen von Ostalpengletschern. *Berichte der Deutschen Botanischen Gesellschaft*, **85**: 1-47.

Patzelt, G., 1973. Die neuzeitlichen Gletscherschwankungen in der Venedigergruppe (Hohe Tauern, Ostalpen). *Zeitschrift für Gletscherkunde und Glazialgeologie*, **9**(1-2): 5-57.

Patzelt, G., 1977. Der zeitliche Ablauf und das Ausmass postglazialer Klimaschwankungen in den Alpen, *Erdwissenschaftliche Forschung*, **13**: 249-259.

Patzelt, G., 1978. Der Österreichische Gletscherkataster. *Almanach '78 der Österreichischen Forschung*, 129-133.

Patzelt, G., 1980. The Austrian glacier inventory: status and first results. *IAHS Publ.*, **126**, 129-133.

Patzelt, G., 1985. The period of glacier advances in the Alps, 1960 to 1985. *Zeitschrift für Gletscherkunde und Glazialgeologie*, **21**: 403 - 407.

Patzelt, G. and S. Bortenschlager, 1973. Die postglazialen Gletscher- und Klimaschwankungen in der Venedigergruppe (Hohe Tauern, Ostalpen). *Zeitschrift für Geomorphologie*, **16**: 25-72.

Paul, F., A. Kääb, H. Rott, A. Shepherd, T. Strozzi and E. Volden, 2009. GlobGlacier: a new ESA project to map the world's glaciers and ice caps from space. *EARSeL eProceedings*, **8**(1): 11-25.

Pellicciotti, F., B. Brock, U. Strasser, P. Burlando, M. Funk and J. Corripio, 2005. An enhanced temperature-index glacier melt model including the shortwave radiation balance: development and testing for Haut Glacier d'Arolla, Switzerland. *Journal of Glaciology*, **51**(175): 573-587.

Pellikka, P. and W.G. Rees, (eds.), 2009. *Remote Sensing of Glaciers. Techniques for Topographic, Spatial and Thematic Mapping of Glaciers*, Taylor and Francis, Lodon, UK, 330pp.

Radić, V. and R. Hock. 2006. Modeling future glacier mass balance and volume changes using ERA-40 reanalysis and climate models: A sensitivity study at Storglaciären, Sweden. *Journal of Geophysical Research*, **111**, F03003.

Rasmussen, L.A. and J.M. Wenger. 2009. Upper-air model of summer balance on Mount Rainier, USA. *Journal of Glaciology*, **55**, 619-624.

Reinwarth, O. and H. Escher-Vetter, 1999. Mass Balance of Vernagtferner, Austria, from 1964/65 to 1996/97: Results for Three Sections and the Entire Glacier. *Geografiska Annaler. Series A, Physical Geography*, **81**(4): 743-751.

Richter, E., 1888. *Die Gletscher der Ostalpen*. Engelhorn, Stuttgart, Germany, 306pp.

Rohling, E.J. and H. Palike, 2005. Centennial-scale climate cooling with a sudden cold event around 8,200 years ago. *Nature*, **434**(7036): 975-979.

Sailer, R., 2001. *Späteiszeitliche Gletscherstände in der Ferwallgruppe*. PhD-Thesis, University of Innsbruck, 205pp.

Schaefli, B., B. Hingray, M. Niggli and A. Musy, 2005. A conceptual glacio-hydrological model for high mountainous catchments. *Hydrology and Earth System Science*, **9**(1): 95-109.

Schiefer, E. and B. Menounos. 2010. Climatic and morphometric controls on the altitudinal range of glaciers, British Columbia, Canada. *The Holocene*, 20(4), 517-523.

Schlüchter, C. and U.E. Joerin, 2004. Alpen ohne Gletscher? *Die Alpen*, **6**: 34-47.

Schöner, W. and R. Böhm. 2007. A statistical mass-balance model for reconstruction of LIA ice mass for glaciers in the European Alps. *Annals of Glaciology*, **46**, 161-169.

Schrott, D. 2006. Flächenhafte Modellierung der Energie- und Massenbilanz am Hintereisferner. (Diploma thesis, University of Innsbruck.), http://imgi.uibk.ac.at/sekretariat/diploma_theses/Schrott_Daniel_2006_Dipl.pdf, 146pp.

Seiser, B., 2010. *Gletscherinventar 2006 der Stubaier Alpen*. Diploma Thesis, University of Innsbruck, 81pp.

Sevruk, B. 2004. *Niederschlag als Wasserkreislaufelement*, Nitra, Zürich, 200 pp.

Singh, P. and V.P. Singh, 2001. *Snow and glacier hydrology*. Kliwer, Dordrecht, The Netherlands, 742pp.

Shea, J., S. Marshall and J. Livingston. 2004. Glacier Distributions and Climate in the Canadian Rockies. *Arctic, Antarctic, and Alpine Research*, **36**(2), 272-279.

Slupetzky, H. 1989. Die Massenbilanzmessreihe vom Stubacher Sonnblickkees 1958/59 bis 1987/88. *Zeitschrift für Gletscherkunde und Glazialgeologie*, **25**(1), 69-89.

Slupetzky, H. 2003. Do we need long term terrestrial glacier mass balance monitoring for the future? *EGS - AGU - EUG Joint Assembly*, Nice, France, 11390.

Solomon, S., D. Qin, M. Manning, Z. Chen, M.C. Marquis, K. Averyt, M. Tignor and H.L. Miller, (eds.), 2007. *Climate Change 2007: The Physical Science Basis. Contribution of Working Group I to the Fourth Assessment Report of the Intergovernmental Panel on Climate Change*, Cambridge,

New York, Intergovernmental Panel on Climate Change, 1009pp.

Span, N., M. Kuhn and H. Schneider 1997. 100 years of ice dynamics of Hintereisferner, Central Alps, Austria, 1894-1994. *Annals of Glaciology*, **24**: 297-302.

Steinacker, R. 1983. Diagnose und Prognose der Schneefallgrenze. *Wetter und Leben*, **35**, 81-90.

Stull, R.B. 2000. *Meteorology for Scientists and Engineers*. 2 ed. Pacific Grove, CA., Brooks/Cole, 502pp.

Suter, P.J., A. Hafner and K. Glauser, 2005. Prähistorische und frühgeschichtliche Funde aus dem Eis - der wiederentdeckte Pass über das Schnidejoch. *Archäologie der Schweiz*, **28**: 16-23.

Thouret, J.C., E. Juvigné, A. Gourgaud, P. Boivin and J. Dávila, 2002. Reconstruction of the AD 1600 Huaynaputina eruption based on the correlation of geologic evidence with early Spanish chronicles. *Journal of Volcanology and Geothermal Research*, **115**(3-4): 529-570.

Uppala, S.M., P.W. KÅllberg, A.J. Simmons, U. Andrae, B. Da Costa, M. Fiorino, J.K. Gibson, J. Haseler, A. Hernandez, G.A. Kelly, X. Li, K. Onogi, S. Saarinen, N. Sokka, R.P. Allan, E. Andersson, K. Arpe, M.A. Balmaseda, A.C.M. Beljaars, L. Van De Berg, J. Bidlot, N. Bormann, S. Caires, F. Chevallier, A. Dethof, M. Dragosavac, M. Fisher, M. Fuentes, S. Hagemann, E. Hólm, B.J. Hoskins, L. Isaksen, P. Janssen, R. Jenne, A.P. McNally, J.F. Mahfouf, J.J. Morcrette, N.A. Rayner, R.W. Saunders, P. Simon, A. Sterl, K.E. Trenberth, A. Untch, D. Vasiljevic, P. Viterbo and J. Woollen. 2005. The ERA-40 re-analysis. *Quarterly Journal of the Royal Meteorological Society*, **131**(612): 2961-3012.

van den Broeke, M., J. Bamber, J. Ettema, E. Rignot, E. Schrama, W.J. van de Berg, E. van Meijgaard, I. Velicogna and B. Wouters, 2009. Partitioning Recent Greenland Mass Loss. *Science*, **326**(5955): 984-986.

van Husen, D., 1987. *Die Ostalpen in den Eiszeiten*. Geologische Bundesanstalt, Vienna, Austria, 24pp.

van Husen, D., 1997. LGM and late-glacial fluctuations in the Eastern Alps. *Quaternary International*, **38-39**: 109-118.

van Husen, D., 2004. Quaternary glaciations in Austria. In Ehlers, J. and P.L. Gibbard, (eds.), *Developments in Quaternary Science*, Elsevier, 1-13.

Velicogna, I., 2009. Increasing rates of ice mass loss from the Greenland and Antarctic ice sheets revealed by GRACE. *Geophys. Res. Lett.*, **36**(19): L19503.

Walcher, J., 1773. *Nachrichten von den Eisbergen in Tyrol*. Kurzböcken, Frankfurt and Leipzig, Germany, 99pp.

Weber, M., 2005. *Mikrometeorologische Prozesse bei der Ablation eines Alpengletschers*, PhD-thesis Innsbruck University, 311pp.

WGMS. 2007. Glacier Mass Balance Bulletin No. 9 (2004-2005), ed. Haeberli, W., M. Zemp and M. Hoelzle. ICSU (FAGS)/IUGG(IACS)/UNEP/UNESCO/WMO, World Glacier Monitoring Service, Zurich, 100pp.

Winkler, S., 1996. *Frührezente und rezente Gletscherschwankungen in Ostalpen und West-/Zentralnorwegen. Ein regionaler Vergleich von Chronologie, Ursachen und glazialmorphologischen Auswirkungen*. Trierer Geographische Studien, **15**, Trier, Germany, 580pp.

Woebcken, W., 1989. 120 Jahre danach - die 1. Bergpanoramen von der Kreuzspitze. *Der Bergsteiger*: 63-70.

Würländer, R. and K. Eder, 1998. Leistungsfähigkeit aktueller photogrammetrischer

Auswertemethoden zum Aufbau eines digitalen Gletscherkatasters. *Zeitschrift für Gletscherkunde und Glazialgeologie*, **35**: 167 - 185.

Yang, D., B.E. Goodison, J.R. Metcalfe, V.S. Golubev, R. Bates, T. Pangburn and C.L. Hanson, 1998. Accuracy of NWS 8" Standard Nonrecording Precipitation Gauge: Results and Application of WMO Intercomparison. *Journal of Atmospheric and Oceanic Technology*, **15**(1): 54-68.

Zachos, J., M. Pagani, L. Sloan, E. Thomas and K. Billups, 2001. Trends, Rhythms, and Aberrations in Global Climate 65 Ma to Present. *Science*, **292**(5517): 686-693.

Zemp, M., W. Haeberli, S. Bajracharya, T.J. Chinn, A.G. Fountain, J.O. Hagen, C. Huggel, A. Kääb, B.P. Kaltenborn, M. Karki, G. Kaser, V.M. Kotlyakov, C. Lambrechts, Z.Q. Li, B.F. Molnia, P. Mool, C. Nellemann, V. Novikov, G.B. Osipova, A. Rivera, B. Shrestha, F. Svoboda, D. G. Tsetkov and T.D. Yao, 2007. Glaciers and ice caps. Part I: Global overview and outlook. Part II: Glacier changes around the world. In *Global outlook for ice & snow*, UNEP/GRID, Arendal, Norway, 115-152.

Zoller, H., 1960. Pollenanalytische Untersuchungen zur Vegetationsgeschichte der insubrischen Schweiz. *Denkschriften der Schweizerischen Naturforschenden Gesellschaft*, **83**: 45-156.

Zoller, H., C. Schindler and F. Röthlisberger, 1966. Postglaziale Gletscherstände und Klimaschwankungen im Gotthardmassiv und Vorderrheingebiet. *Verhandulungen der Naturforschungsgesellschaft Basel*, **77**: 97-164.

ACKNOWLEDGEMENTS

At the end of this work I feel the necessity to thank several persons and institutions for the support I could count on in many different ways:

I am very thankful to my advisor Michael Kuhn. He was a great teacher for me in the past years. He encouraged me scientifically and showed me a straight way to pursue my ideas. He always had an open door, never gave me the impression of disturbing him, which, together with his enthusiasm for science will always remain an ideal for me. For the liberty he gave me, for the support to follow side-projects that are very beneficial to me but not directly related to my work but above all, for the modesty and generosity he showed me, I express my sincere thanks.

My family has been a big moral support in the past years. Thanks for believing in me, embedding my work into such positive surroundings and giving me the liberty I need.

Thanks to my colleagues at the Insitute of Meteorology and Geophysics for great times scientifically, logistically and personally, many among them became friends in the past years.

Thanks to the Commission for Geophysical Research of the Austrian Academy of Sciences for funding and their employees for very non-bureaucratic collaboration.

Thanks to the University of Innsbruck, especially to the Institute of Meteorology and Geophysics for providing me with the necessary working environment and logistical support.

While writing these lines, I started naming people who contributed to filling these years with positive memories and helped me in many different ways. I soon realized that it would make this part way too long and, more importantly, the danger of omitting someone or putting into wrong orders is too high. I therefore do not name any of them (my sincere apologies for that!) here but I hope and actually I am confident that they know their position and what I owe them. Thank you!

Die VDM Verlagsservicegesellschaft sucht für wissenschaftliche Verlage abgeschlossene und herausragende

Dissertationen, Habilitationen, Diplomarbeiten, Master Theses, Magisterarbeiten usw.

für die kostenlose Publikation als Fachbuch.

Sie verfügen über eine Arbeit, die hohen inhaltlichen und formalen Ansprüchen genügt, und haben Interesse an einer honorarvergüteten Publikation?

Dann senden Sie bitte erste Informationen über sich und Ihre Arbeit per Email an *info@vdm-vsg.de*.

Sie erhalten kurzfristig unser Feedback!

VDM Verlagsservicegesellschaft mbH
Dudweiler Landstr. 99
D - 66123 Saarbrücken

Telefon +49 681 3720 174
Fax +49 681 3720 1749

www.vdm-vsg.de

Die VDM Verlagsservicegesellschaft mbH vertritt

Printed by Books on Demand GmbH, Norderstedt / Germany